SUCCESSFUL
COST REDUCTION
PROGRAMS FOR
ENGINEERS
AND MANAGERS

SUCCESSFUL COST REDUCTION PROGRAMS FOR ENGINEERS AND MANAGERS

E. A. Criner, P.E.

VNR VAN NOSTRAND REINHOLD COMPANY
NEW YORK   CINCINNATI   TORONTO   LONDON   MELBOURNE

Copyright © 1984 by Van Nostrand Reinhold Company Inc.

Library of Congress Catalog Card Number: 83-5787
ISBN: 0-442-21579-7

Manufactured in the United States of America

Published by Van Nostrand Reinhold Company Inc.
135 West 50th Street
New York, New York 10020

Van Nostrand Reinhold Company Limited
Molly Millars Lane
Wokingham, Berkshire RG11 2PY, England

Van Nostrand Reinhold
480 Latrobe Street
Melbourne, Victoria 3000, Australia

Macmillan of Canada
Division of Gage Publishing Limited
164 Commander Boulevard
Agincourt, Ontario MIS 3C7, Canada

15 14 13 12 11 10 9 8 7 6 5 4 3 2 1

Library of Congress Cataloging in Publication Data

Criner, E. A.
    Successful cost reduction programs for engineers
and managers.

    Includes Index.
    1. Engineering economy.   2. Cost control.
3. Costs, Industrial.   I. Title.
TA177.7.C74 1983   658.5   83-5787
ISBN 0-442-21579-7

Preface

Successful Cost Reduction Programs for Engineers and Managers is written for engineers and managers who are concerned about today's competitive environment. In particular, it emphasizes useful fundamentals and illustrates numerous cost reduction ideas that have application to a broad range of production, warehouse operations, and associated expenses. The information provided covers every major aspect in support of a successful cost reduction program.

Today's engineers and managers need to cope with a dynamic technology and variable business conditions to remain competitive with internal as well as external sources. This book will aid you in important decision making by illustrating why, where, and how cost reduction ideas can be initiated, developed, and processed to a successful conclusion. Direct labor savings and material savings coupled with effective expense control will continue to be the most fruitful results of a successful program.

For engineers, cost reduction serves as a standard by which professional achievement and performance appraisals are measured. Involvement with a successful program provides personal recognition and job satisfaction, affects salary administration, and also impacts on consideration for promotion. For managers, cost reduction is a key ingredient in producing a quality product on schedule at a competitive cost. Success in these areas can be linked to monetary rewards and upward mobility in the management structure.

Whether your interest is in developing a new cost reduction program or improving an existing program, this book will provide the facts to achieve positive results.

Although a major portion of my experience has been with Western Electric Company and a number of suppliers that serve the Bell System, I have selected and presented a wide cross section of cost reduction applications. My intent has been to utilize as many examples as possible outside the Western Electric environment. Meaningful material pertaining to Western Electric has been retained to illustrate general as well as specific applications. Specifically, my objective has been to provide the reader with as many varied solutions to cost reduction program problems as possible.

I would like to express my appreciation to Western Electric Management for their support in making this book possible. This includes their sharing of data and help in reviewing the book on a chapter by chapter basis. I also would like to express appreciation to the editorial staff of Van Nostrand Reinhold Company, for their courteous cooperation in the production of this book and, above all, to my wife, for her cheerful encouragement in spite of the demands made by typing the manuscript in conjunction with her regular full-time job.

Finally, I would like to express my appreciation and indebtedness to my fellow engineers who freely shared information related to their accomplishments, to the American Institute of Industrial Engineers, and the American Productivity Center for permission to reproduce their material as acknowledged in the text.

The concerns and involvement relating to cost reduction as a stepping stone to increased productivity will continue to experience significant changes in the 1980s. It is stimulating to participate in these developments. Hopefully the material in "Successful Cost Reduction Programs" will contribute to a better understanding of the theory and practice and the relationship to productivity.

E. A. Criner, P. E.
San Jose, California

How This Book Will Help You

This book presents in a logical sequence the facts that are associated with a successful cost reduction program.

Cost reduction first starts with an idea, an investigation, or an analysis of an ongoing function. The techniques presented will enable you to identify areas that have cost reduction potential. Examples and illustrations of a wide variety of cost reduction cases are presented.

Examples of Savings:

Project	*Annual Savings*
• A large California winery installs a co-generation system in order to reduce costs.	$125,000
• A large manufacturing company takes a new look at packaging circuit boards. The result is increased productivity and reduced costs.	$325,000
• A company takes a second look at the disposal of waste packaging materials. As a result savings are now being achieved in disposal charges and the consumption of oil and gas.	$104,000
• The addition of automatic tape labeling machines in a warehouse operation saves both labor and material.	$77,500
• The use of a robot for making welds is introduced into a shop operation.	$12,000
• A change in procedure for receiving material in a distribution network provides a more cost-effective operation.	$154,300

- A programmable lighting control system is proven to be an effective cost containment tool. — $130,000
- A large bank analyzes the procedures associated with the processing of checks. — $703,000
- A survey of standard forms produces savings in labor, storage charges, and the total number of forms that were stocked. — $102,000
- A warehouse consolidation produces savings in labor, investment, and related operating costs. — $192,000
- An updated lighting system involving the replacement of 400 watt mercury-vapor lamps with high-pressure sodium lamps provides substantial savings. — $48,000
- The introduction of new laser soldering techniques are used to replace a manual system. Substantial savings result from the process change. — $55,200

Project	*Annual Savings*
A vibratory finishing system is installed to increase the recovery percentage in the painting of plastic parts.	$71,900
An engineer conducts a study to determine the proper size of supply fan motors for air make-up systems. The result is reduced energy consumption.	$14,800
A change in packaging procedures eliminates the use of special materials. The change to a tray-pack system is introduced.	$34,700
A large regional office introduces a revised specification for computer	$125,000

paper. A change in size and weight of
the paper provides substantial savings.

Other case studies are included in the book. The cases and cost reduction briefs cover a wide spectrum of activity from many industry
segments. The applications have one common link—that a change or
improvement was introduced into a process, a product, or a service.
As you know, changes do not just occur; they come about by effective
actions that are initiated by engineers and managers.

Experience has proven that controlled change through an effective
cost reduction program contains the necessary checks and balances to
answer the following questions and others that may arise.

- Is the idea or proposal cost effective?
- Does the rate of return justify the capital expenditure?
- Does the change have any positive or negative impact on the
  product or service?

In summary, the effective results generated from an ongoing cost
reduction program can be traced through the accounting maze and
measured as increased productivity at the firm level.

Contents

SUCCESSFUL COST REDUCTION PROGRAMS FOR ENGINEERS AND MANAGERS

1
Introduction

An ongoing cost reduction program is an effective tool that must be in daily use in all companies, both large and small, if they are to survive in the 1980s. The installation of a logical system of controls as a necessary overall procedure in manufacturing, transportation, trade, and finance is a sound method that leads to cost reduction in most applications. *Control* is a strong word with certain ominous implications, and in most areas would imply a restriction on involvement and freedom of action. However, this is not true in the operation of most cost reduction programs.

WHO NEEDS A COST REDUCTION PROGRAM?

Who needs a cost reduction program in today's business environment? The answer is any organization that expects to survive inflation, the quality decline, and the productivity decline that has touched almost every segment of the economy.

In today's economy and the economy of the future, we will survive and prosper only if we, as managers, engineers, and others, can achieve a firm grasp on the major cost inputs in our business. In general terms these inputs are defined as labor, material, energy, capital, and related expense. Efficient control of these key inputs can be traced through the cycle and measured as outputs that can be realized in the final analysis.

Cost reduction development in a large company may encompass teams of engineers in different disciplines working toward a common goal of producing a product or service in a more efficient manner. In a small organization the total formal effort of reducing costs may be handled by one person, perhaps on even a part-time basis. Progress in this endeavor has developed in modern industry largely as the result of the ever-increasing demands made on manufacturing facilities. These facilities are expensive, and in order to generate a satisfactory profit it is necessary to plan and control the flow of materials to obtain the lowest overall costs. Engineering design and specification create new products in order to be competitive in the consumer market. Modifications are made to update the product as the need for better performance and competition dictates. However, even if no changes in design, specifications, or methods were made, after the product was placed on the market it would still be necessary to plan and schedule ways to reduce costs.

THE APPROACH

The cost reduction approach can be defined as an overall philosophy that blends the various engineering decisions associated with labor, materials, energy, capital, and related expense. The manager and engineer in all segments of industry have an obligation to insure that only the required amounts of the cost inputs listed above are used to produce any product or service. How many people should be assigned in the pursuit of cost reduction is a question that is often asked. In general terms the answer is based on the long- and short-term estimates of future sales for the total company or specific product lines in a segment of the company. The application of effort can in part be determined by the profit or loss from specific items, or expanded to cover a total plant operation or the overall operation of the company.

This book is primarily concerned with the development, investigation, processing, and introduction of cost reducing ideas into the production process. Updating the progress that is being made is essential for informing executives in engineering, production, sales, accounting, and other management areas. The paperwork that conveys this information may vary widely from one industry to another, but this is not important in itself if the system selected works and produces positive results that can be measured and credited to the individual responsible for suggesting the change.

THE FOCAL POINT: COMPETITION

It is becoming increasingly clear that the United States is locked in a ferocious economic competition with other industrialized countries such as Germany, France, Italy, and Japan. The competition is taking place in a number of industrial areas, including automobiles, cameras, consumer electronics, and sophisticated telephone communications systems.

The focal point of this competition is the increasingly slow rate of growth of productivity in the industrial sector. Productivity is defined as goods or services produced per man-hour worked. As productivity has slowed in the United States, the common presumption has been that lack of efficiency in our factories is the culprit.

PRODUCTION EFFORT VERSUS EXPENSE OPERATIONS

A number of articles have recently been written dealing with use of industrial robots in Japan. Some would even say that world leadership in the development and use of such robots is now held by Japan. At the same time much has been written about the virtues of Japanese workers, comparing them favorably with American workers. Are we looking at only part of the productivity problem? The answer to this question appears to be yes. The ratio of workers in the actual manufacturing process compared to support activities has shifted in recent years.

A few years ago, based on data issued by the Bureau of Labor Statistics, approximately 70 man-hours of each 100 hours worked in a manufacturing company were devoted to working on and producing the products. Thirty man-hours were spent in support of manufacturing activities such as accounting, purchasing, data processing, and other related functions. Today the process has shifted, with even a higher percentage—perhaps in the area of 50 percent—going to related paperwork oriented to support the manufacturing process. What is being done to improve the productivity of the growing administrative labor force? In the typical company, very little.

Production work has been studied, measured, and monitored carefully in every industry. This is easy to understand since this type of involvement can be easily measured and improved. However, we do not really know what clerks, accounting people, or even managers do, how well they do it against a given standard, or how well they do it in comparison to their peers. In other words, we do not measure administrative work. The question then is, if we do not measure it, how can we hope to make it more productive?

A VIEW OF THE REMAINING CHAPTERS

From the discussion thus far it is easy to see that the concept of productivity has been misunderstood. In future chapters we will deal with cost reduction as a tool for reducing the input equation for productivity. In summary, we will deal with cost reduction as a stepping stone to productivity in the 1980s.

In spite of the increased attention that has been paid to cost reduction in recent years, there are still widespread misunderstandings concerning the nature and the meaning of cost reduction concepts. The important thing is that the findings of various cost reduction approaches should be assembled in a systematic form and made available to the public in general. Only by increasing our knowledge of the cost reduction approach can we work toward improving productivity and maintaining a competitive position in the United States and world markets.

The intent of this book is to summarize the current state of knowledge concerning cost reduction, with specific references and illustrations that show the workings of an effective program. It is the aim of the author to show the development, operation, and results reporting that are essential for a successful program. In order to do this we must develop new and improved modes of production, applying new technologies, creating, adopting, or adapting work methods, processes, and new concepts in order to obtain increased output for the same input costs. Achievement of this goal requires involvement that is creative, imaginative, and well documented in order to protect the originator of the idea. Effective managers structure organizations that generate creative ideas from employees at all levels and involve them in the innovation and implementation associated with technological progress.

In Chapter 2, a review of the origins of cost reduction is presented. Current challenges that we now face are discussed. In order to develop an appreciation for the problem we must take a new look at the productivity. A brief summary of the Western Electric Cost Reduction Program is discussed. Tips on getting involved in cost reduction are covered, and results from a successful program are reviewed. Examples of cost effective products and services are cited in order to pull the whole concept together.

The major benefits of an effective cost reduction program are detailed in Chapter 3. Cost reduction potential is reviewed in the area of energy and in other fields. Feedback from a large management consulting firm is shared. The concept of cost reduction versus cost avoidance is covered. Comments from a comprehensive audit are reviewed.

Chapter 4 deals with how to get started. Each job assignment has a certain amount of cost reduction potential. An overview of cost reduction opportunities is reviewed. Development of ideas is the real challenge that leads to change and improvement. Feedback from an industrial engineering consultant is shared. A discussion dealing with problem identification concludes the chapter.

Organizing for cost reduction is reviewed in Chapter 5. The puzzle aspects of labor, material, energy, scrap, capital, and expense are covered. Cost reduction examples are cited that illustrate typical applications. Preparation of the required paperwork is included. The key question, if you do not have an ongoing program, is how to start one. Problems and challenges associated with your program will vary from one industry segment to another.

Chapter 6 deals with setting target goals. For the sake of the company and yourself, don't be afraid to rock the boat. Remember that cost reduction is everybody's business. Results from various programs will vary, but the main element is idea input and continued involvement. Time management is the key to real effectiveness. Record ideas as they occur and plan a follow-up schedule. The case study from General Foods provides an excellent example.

Monthly progress meetings are discussed in Chapter 7. These formal meetings are necessary to deal with cost reduction cases in an effective manner. As with any organized involvement—baseball, football, basketball, and cost reduction—people play certain roles. Guidelines for conducting meetings are reviewed, and a sample set of minutes that reflect the achievements is presented. A typical cost reduction flowchart is illustrated. Productivity considerations are highlighted, and measurement of current objectives and future goals is covered.

Chapter 8 deals with expense control techniques. Employees play a major role—the input of new ideas is a must. Each employee must be encouraged to think, write, and submit his or her ideas. Increased productivity does not just happen. Dr. Frank E. Cotton, Jr., shares some vital comments on this subject. A case study dealing with the measurement of productivity at United Airlines concludes the chapter.

Potential savings associated with shop operations are covered in Chapter 9. Challenges and opportunities are outlined. Cost control concepts that keep management and engineers informed are discussed. Techniques on reviewing results are cited, and common manufacturing problems are listed and commented on. The need for long-range planning is discussed and planning factors are illustrated. The chapter is concluded with a look at ways to identify productivity opportunities. Ruddell Reed, Jr., shares some interesting observations and concepts that have a wide range of applications in the shop, warehouse, and office.

Chapter 10 deals with warehouse operations. In general terms, the warehouse function can be defined as a nonvalue-added operation. In order to secure a firm grasp on this expense function, an in-depth analysis is needed to examine the cost of space, labor, facilities, and other expense items. This will establish a basis from which real progress can be made. Labor utilization is the cost component that offers the biggest challenge. How can warehouse operations be compared one against the other? The best way to answer that question is to conduct your own study. A list of partial productivity measures is provided for your consideration. A computer space allocation program can be useful when you are reviewing the need for space requirements in terms of bin storage and pallet racks.

In Chapter 11 the relationships between employees, expenses, and expectations are examined. The productivity challenge can be met by reducing input costs while holding the same rate of output. Greater management involvement is the key to productivity improvement. Work relationships and effective feedback in both directions are essential. The first-line supervisor is the key in dealing with all employees. Comments on how to become a productivity supervisor are shared for your review. Expense control techniques that may apply in your environment are listed. One major segment of the workforce that is often overlooked is the clerical staff. R. Keith Martin shares some ideas on how to look into this important segment.

Cost reduction in the public sector is examined in Chapter 12. There have been many misconceptions about the applications of proven cost reduction techniques in various levels of the public sector. These long-held beliefs may be changing. The passage of Proposition 13 in California in 1978 was a major force in this change. The people by their own vote created, by popular demand, one of the largest cost reductions on record. Property taxes were slashed by almost 60 percent—approximately $7.0 billion. A cross section of expense-related problems that confront other segments in various parts of the country are examined.

Industrial engineering techniques are being utilized in Winston-Salem, North Carolina. This initial application of engineering to public sector problems was introduced in the early 1960s. A typical study format that was developed by S. H. Owen, Jr., is included for possible application. Also included is a paper authored by Richard L. Shell and Dean S. Shupe. A wide number of problem areas are discussed.

Chapter 13 deals with conducting cost reduction cases. The total impact of cost reduction is discussed. The buyer of a product or service, the employee, and the stockholder each have a different view of the impact of cost reduction. Ten items for cost reduction consideration are listed. This quest for effective cost reduction cannot be viewed as a temporary involvement; it must become a full-time involvement. Verification of results by accounting is an ongoing requirement. The approach may be a concerted effort to examine items that in the past have been accepted as the status quo. Some may think that cost reduction is a tool for use in big companies only. This is not so: It also has application in other environments. Data from Central Methodist College, Fayette, Missouri, is shared for your review. In summary, cost reduction is related to problem solving. Some problems are known, others are unknown.

Chapter 14 deals with cost reduction results. In order to achieve results, the function must be ongoing—not a start-and-stop involvement. A cross section of cases from Western Electric and other sources are discussed. A point to remember is that these achievements did not just happen; they were the result of carefully planned engineering investigations. The applications, while quite varied, have one common bond: Each was directed at producing a quality product in a more cost-effective manner. A case study from Paul Masson Vineyards Winery concludes the chapter.

Getting into the act is the real challenge. Chapter 15 deals with various aspects of getting involved in the program. Once involved in the program, it is possible to turn these results into a forecasting tool that can be used by management. Feedback on the status of each cost reduction case is a must. Monthly feedback is essential in order for the program coordinators to maintain adequate control. Problems encountered in the quest for cost reduction should be met in a straightforward manner. No one ever said that obtaining results would be easy. Details on a plastic recycle program are reviewed to illustrate a point.

Where do you go from here? Bob Zingali, cost reduction coordinator at Merrimack Valley, provides some interesting comments. A checklist for planning cost reduction effort concludes the chapter.

Chapter 16 provides a summary of the total involvement. The acid test of any program is evaluation of the long-term trends in price, quality, and service. It would be a positive step to find something that could be used as a standard reference of days gone by that is still a good deal today. Such a product does exist and we all use it daily. The product is the telephone and the wide range of related services that we have all grown very dependent upon.

What leads me to think this way in today's environment? My decision is based on three factors: price, quality, and service.

A question that is often asked is, "Can the quest for continued cost reduction be carried too far?" The answer is yes. As a case in point, let us look at a report on a symphony orchestra.

REPORT ON A SYMPHONY ORCHESTRA

The following article purports to be a report by an engineering procedure examiner after attending a symphony concert. It will strike a sympathetic chord among those supervisors who have experienced the results of a visit by one of these experts and *their* opinion of what the supervisor considered a well-organized and operating setup.

> For considerable periods the oboe players had nothing to do. The number should be reduced, and the work spread out more evenly over the whole of the concert, thus eliminating peaks of activity.
>
> It is noted that all twelve first violins were playing identical notes. This seems unnecessary duplication; the staff of that section should be drastically cut. If a large volume of sound is required, it could be obtained by means of electronic-amplifier apparatus.
>
> Much effort was absorbed in the playing of sixteenth and so-called grace notes. This is an excessive refinement. It is recommended that all notes should be rounded up to the nearest eighth note. If this were done, it

would be possible to use trainee and lower-grade operatives more exclusively.

There is too much repetition of some musical passages. Scores should be drastically pruned. No useful purpose is served by repeating on horns and woodwinds a passage which has already been adequately handled by the strings. It is also estimated that if all redundant passages were eliminated, the whole concert time could be reduced to 20 minutes, and there would be no need for an interval or intermission. The conductor concurs generally with these recommendations, but expresses the opinion that there might be some falling off in box office receipts. In that unlikely event, it should be possible to close sections of the auditorium entirely, with a consequent saving in overhead: lighting, janitor, service, heating, etc.[1]

Is it possible for the concepts associated with a cost reduction program to be carried to extremes? The answer is yes. The benefits from a cost reduction approach are lost if the output of the service or product are altered to such a degree that the attributes of price, service, or quality are altered.

REFERENCE DATA

1. Kamman, H. E., Consulting Engineer, San Jose, Ca.

2
Background

Today our nation's economy stands at a crossroads. The issues that must be dealt with are inflation, energy dependence, high capital costs, and in summary, a slowed economic environment. The engineer and manager in industry and other segments of the economy are being challenged as never before. High labor costs, increasing material costs, and declining quality are only a few issues that must be addressed. Until 1970, productivity growth in the private sector of the U.S. economy averaged over 3 percent a year, according to the Bureau of Labor Statistics. After 1970, the trend line for productivity has been downward, dropping to 1.8 percent for the period 1970–1978.[1]

WHAT IS PRODUCTIVITY?

In order to define the problem, let us make sure that we understand certain aspects that do or do not have an impact on productivity.

Productivity is not just producing more units.
Productivity is not solely directed at the production worker.
Productivity does apply to the prudent use of our resources that are required to generate a given product or service.
Productivity does require the constant effort and ingenuity of engineering and management.

In summary, productivity can be defined as the output over the input; the relationship between the total physical output of a production shop, a factory, or a complete industry segment, and the factors of input, labor, capital, energy, material, and related expense.

$$\text{Total Productivity} = \frac{\text{Output}}{\text{Input}}$$

CONTROL OF THE EQUATION

In order to improve productivity it is essential that we, in fact, obtain better control of the input equation. We must improve our ability to control labor, capital, energy, material, and related expense.

Putting the problem in perspective, the productivity continues to fall. Stated again: Productivity is a measure of goods and services that the economy produces per hour of paid working time. The continued decline in productivity has alarmed government economists because this means rising unit labor costs that contribute to high inflation.

Productivity Increases in Manufacturing, 1968—1978.

COUNTRY	10-YEAR DATA	AVERAGE YEAR
Japan	89.1%	8.91%
Germany	63.8%	6.38%
France	61.8%	6.18%
Italy	60.1%	6.01%
United States	23.6%	2.36%
United Kingdom	21.6%	2.16%
(Data from the Bureau of Labor Statistics)		

These data are shocking, but they are revealing, too. America has given superb instruction in the art of industrial excellence. It is clear, however, that some of the lessons have been lost to the teacher. Products once poured out of our factories to set worldwide standards of quality: steel, automobiles, production machinery, and a wide range of electrical products.

Now products flow into America while we face economic stagnation and heavy trade deficits. This situation could get worse before we make any real progress. Other countries have hit their industrial stride and are likely to pull ahead even further unless we do something. American industry can reverse these productivity trends.

COST REDUCTION, A NEEDED TOOL

Cost reduction can be one of our most important tools in changing these trends. An effective program deals with labor, materials, energy, capital, and other related expenses. An effective cost reduction program has three major areas of benefit. Let us examine each of these areas—company benefits, employee benefits, and customer benefits.

Company benefits can be in the following areas: increased efficiency, reduced production cost, improved quality, and improved productivity. These items are very important in keeping pace with competition.

Employee benefits for the engineer or manager are also important. Involvement in cost reduction provides personal recognition and job satisfaction, affects salary administration, and also impacts on consideration for promotion.

Customer benefits are where the total impact of cost reduction is evident. The desired end result is in the form of a reduced price, with no compromise in quality, and reduced lead time to obtain the product.

Cost reduction is a team effort that involves the company, the employee, and the customer. In broad terms cost reduction deals with a reduction of labor, material, capital, energy, floor space, and expense costs. This holds true in a small company as well as a large one.

Labor cost can be reduced by changing methods, improving material flow, and redesigning tools and equipment. Are you as an engineer or manager satisfied with the progress in your area of responsibility?

Material cost can be reduced by changing the design of the product, eliminating unnecessary components, and replacing expensive materials with more economical ones. You have probably heard this before. The real question is, how can your progress in this area be improved?

Energy cost can be reduced by planned periodic energy audits in all phases of the operation. A good place to start is with your last utility bill. Does your company have an ongoing energy management program? If there is no program, how can one be implemented?

Floor space can be reduced by rearranging the work areas in a more functional manner, using a two-shift operation where possible, and using vertical air rights where possible. Take a look at the current layout. Analyze the flow of materials and look for ways to improve the operation.

Capital cost can be reduced by—economic justification on each proposal and a priority need index. One key question, is what are the requirements concerning return on investment? The R.O.I. can be used as a tool in selecting the most feasible applications that are related to capital expenditures associated with cost reduction proposals.

Expense costs can be reduced by careful review of support functions such as security contracts, control of expense supplies, janitorial contracts, and other related services. A close examination of these items can be very rewarding.

COST REDUCTION APPLICATION

Cost reduction techniques can be utilized in operations associated with any company. This applies to small concerns as well as companies with multiplant operations. The opportunity to participate in the cost reduction quest is limited only by your imagination. Some job assignments will by their nature present unique challenges and the chance to see the result of your personal involvement.

Personal involvement and the concern for constructive change can be the difference between an outstanding cost reduction program and a program that just moves along and needs constant prodding by management to keep it alive. In today's competitive economy few people would deny the need for constant involvement in containing costs. However, it is not uncommon for this concern to be expressed in a loose, unorganized manner. Results from this approach are sometimes less than acceptable.

There have been many misconceptions dealing with the desired end use associated with the cost reduction results. When applied properly, a cost reduction program will produce results in the form of dollar savings. These savings can in turn be traced through the various inputs of the productivity equation. Cost reduction results should be viewed as a stepping-stone to increased productivity.

INVOLVEMENT

The size of the organization that you as an engineer or manager work in will dictate the structure of the cost reduction organization. A small company can maintain an effective effort and still keep a simple structure. Two essential ingredients are goal setting and the evaluation of input ideas that can lead to the achievement of the desired goals. In general, the smaller company can evaluate and install cost reduction procedures in a shorter time frame and still achieve positive results.

Achievement goals are a must. To reach these goals a plan of action is required. A normal flow in goal setting is from the top of the organization downward. Once the goals have been agreed upon, the challenge begins.

A typical list of goals for a manufacturing-oriented organization could be expressed in the following manner:

- The owner or company president desires to increase total profits by 10 percent.
- Production desires to reduce planned operating costs by gaining better control of the major inputs—labor, material, energy consumption.
- The sales and marketing force have a goal to increase sales by 20 percent with a 10-percent decrease of expense in the achievement of their goal.
- Overhead expense reduction goals for the staff and support group have been established at a 15-percent level.

After these goals have been established and agreed upon, an action plan is required to measure progress toward the goals. The approach will be different for each type and size of company. One thing will be common. The areas to look for savings in will be labor, material, capital, energy, and related expense.

HOW TO GET STARTED

Getting started is the challenge. Everyone has a technique that works best for him or her when it comes to developing new ideas. Let me share with you some techniques that have worked for me. How do you set the stage to develop an idea that may develop into a *creative cost reduction?* Each individual may have a unique plan that works for him or her. If you want to be an idea person, school yourself in these fundamentals.

1. *Think the "green-light, red-light" way.* Suppose your problem is lack of time to train workers. First, apply freewheeling (green-light) thinking to turn up all the solutions: You can assign more workers, work longer, train on overtime, speed up training. Don't stop with less than 15. Then switch on your judgment (red-light thinking), and pick the best. Try to alternate your

green- and red-light thinking at every stage from stating of the problem to final "selling" of the answer.

2. *Narrow down the problem.* John Dewey, great American philosopher, said, "A problem well stated is half solved." This is just as true today. If you don't know that you have a problem or an opportunity, how can you proceed in a positive manner?

3. *Concentrate.* If you worry over a hat full of problems all at once, you'll get flabby ideas. So put on your mental blinders. Concentrate on one—and don't let anything distract you. Make notes and jot down questions that can be answered later.

4. *Keep plugging.* When you draw a blank, it's hard to keep on thinking. A great rowing coach said, "If you can hold on just two strokes longer than your opponents, you'll win." This is also true in the quest for cost containment, cost reduction, or whatever you may call your program.

5. *Believe in yourself.* You can come up with good ideas. And you can do it in one or more of three ways: imagination (weaving ideas into new combinations while thinking deliberately), inspiration (suddenly creating ideas automatically from chance observation or circumstance), and illumination (letting ideas come out of nowhere, when you least expect them—after thinking has stopped). Keep good records of ideas and new concepts as they develop. Review these notes as time permits.

6. *Let your unconscious take over.* When you're worn out, stop thinking. Your mind will keep on working even if you aren't consciously thinking. It will percolate fresh ideas when you return to your problem. If you have doubts on this technique, give it a try and see what happens.

7. *Keep 'em flowing.* If your mind is hot, keep thinking up ideas. If you stop and then try to start again later, you may lose vital thoughts. Some people seem to have time slots when they can be more creative. Ask yourself what time of day or night is best for you.

8. *Act.* Idea creation begins with the hot flash, but it's not completed until you put your idea to work. Crude preparations of penicillin were described in 1929, but nobody followed through on the discovery for a dozen years. Simply stated another way, when you have an idea, don't put off the introduction. Act while you are motivated.

HOW AN EFFECTIVE PROGRAM WORKS

There are no guaranteed standard approaches in the quest for cost reduction. The program must be tailored to the specific company, large or small. The flow of ideas and related paperwork can be adapted to any economic environment.

Since a large part of my experience has been with Western Electric Company, let me share some of my experiences relating to the ongoing program in a large company that has a multiplant operation. Once we forget about the company size, we are back to the basics—ideas, people, follow-up, and effective change that can be translated into cost savings.

Engineers at all Western Electric locations can tap vital company resources for data related to their specific projects. In general, good engineering—an evaluation of new materials, new methods, new developments, and new skills at the right time—characterizes most significant cost reductions. Also, other factors are: good engineering supervision, efficient group effort spearheaded by the product engineer, and the cost-conscious industrial engineer.

Conducting a cost reduction case is done in three distinct stages—opening the case, conducting the case, and closing the case.

Opening the Case

After a review of the considerations listed above, the engineer documents his or her idea in an opening statement that is forwarded to the local cost reduction committee for review and approval. If the idea is beyond the scope of the local committee, all related paperwork is forwarded to the appropriate location for review and approval. After the required approvals to open the case have been secured, the case is routed to accounting for checking and classification.

At this point, the engineer is free to proceed with the development of the idea. In a smaller company the case opening could be accomplished in a discussion on the production floor. However, approvals for expending company funds could lead to a changed course in the project or perhaps see the project halted.

Conducting the Case

In the conducting phase, the engineer has the opportunity to combine "blue-sky concepts" with proven engineering principles. Development expenditures for equipment, design, and other related expense are made against the approved cost reduction case. Cases can cover the gamut from a complete product design to a change in the packing carton or mode of shipment to the end-use location. Lapsed time in this phase can vary from several months to as much as five years or longer depending on the scope of the case. Collecting the data and proving the idea comprise the challenge confronting each engineer involved in this endeavor.

Closing the Case

After the case has been investigated and the method of change has been implemented or installed, the case is prepared for closing. At this time a total economic package is presented to the local cost reduction committee. Savings are expressed on a current level basis (the first year) and also on a five-year summary average.

Your Program

Your program may be quite different due to size, management, your involvement, and a number of other different factors. The only important thing is that your program stimulate ideas, lead to constructive change, and produce cost savings. These items are a must for any program in either a small or large company.

Employee Recognition

In progressive companies management has seen the need to develop recognition programs in conjunction with the cost reduction program. A program of recognition for cost reduction achievement is worthwhile, although some may say that it is really not needed. After all, this is the only reason that engineers are on the payroll. If we were to examine this further, we might conclude that managers were paid to manage. If this were the case, why should they be paid a big bonus if they manage in an effective manner?

The use of a recognition program is not something that has to be introduced at the start of the program. Such an enhancement could be introduced at any time that management deems appropriate. What should such a program contain? That depends on what the management feels is appropriate.

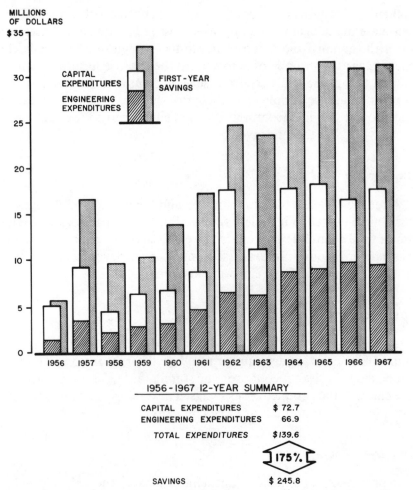

Figure 2-1.   First year savings have more than recovered cost reduction expenditures. (*Source:* Western Electric Company, Inc.)

One progressive company in the Bay Area has recently embarked upon a new program for employee recognition. Highlights of the program are:

1. Publicity for adopted cost reductions in local newspapers and regional news media.
2. Tangible rewards for cost reduction originators. These can range from a $25 gift certificate for cases with savings in the $5,000–

$25,000 range, and a $50 gift certificate for savings in the range of $25,000–$50,000. A gift certificate for $100 is awarded for savings of more than $50,000. The above awards can be presented at the close of the cost reduction year. As an ongoing program, awards are made for cumulative savings on an individual basis.

A Unique Application

The ultimate in cost reduction application is achieved when something that was previously junked, discarded, or thrown away can be reused or recycled.

In 1980, a small San Jose firm, Zerpa Industries, Inc., was geared up to produce an energy-efficient heat pump. Delays in research funding from the U.S. Energy Department led the company into reviewing other products that may have application in the Silicon Valley. As a result of these investigations Zerpa started working on another project, a system to reduce costs by recycling the chemical solvents used by Touche Manufacturing Company. The developed recycling system, known as Zerpa's Recyclene R-14, has evolved to fill a real need.

The recycling system that started as an experimental item has become a hot item on the market. Today the firm has over 1000 back orders for a unit that retails for approximately $3950. Touche, a manufacturer of computer cabinets, was confronted by the rising expense of chemical solvents. Commercial recycling was available, but only for large volumes of solvent. The need that Touche was trying to satisfy dealt with only a few gallons a week.

The new system introduced by Zerpa can recycle up to 95 percent of the used solvent. It can process 13 gallons of chemical waste in approximately ten hours. Recycling eliminates the need to have the solvent shipped to an authorized dump site. Strict cradle-to-grave regulations dealing with hazardous wastes have pushed the cost of disposing of associated chemicals to between $40 and $75 per 55-gallon drum, according to a Zerpa source. Users can gain substantial savings by recycling their solvents instead of buying new. General-use solvents cost between $5.50 and $9.00 per gallon. Zerpa says its recycle system can reclaim the solvent for about ten cents per gallon.

The Recyclene unit is a stand-alone sink with a drum attached to the bottom. The chemical solvent comes out of a tap that resembles a water faucet. Once the solvent has been used it is poured down the drain, which conducts the solvent into the drum. The drum is lined with a nylon bag, which is used to remove the contaminants after distillation of the solvent is completed. When the drum is full the unit is switched on. For eight to ten hours the solvent is heated to its condensation point. The fumes are carried out of the drum and into a holding tank contained within the unit.

When the process is completed, all that is left in the drum is the waste material at the bottom of the heat-resistant nylon bag. The bag serves a function similar to a trash can liner. In the holding tank, clean solvent is ready for use again via the tap.

How about cost savings? A company with a daily usage of ten gallons of solvent could save more than $15,000 yearly. An isolated example of cost reduction you may ask? Not so. This is a new opportunity that did not exist for small solvent users before.

Summary

With any cost control program such as cost reduction, value analysis, cost avoidance, or an employee suggestion system, there are a number of avenues that can be explored.

As approaches are developed to gain control of rising costs, the approaches will deal with the inputs of the productivity equation. These inputs are labor, materials, capital, energy, and expense. No organization is too small or too large that it can afford to overlook potential savings that can be achieved through an organized effort to control and reduce expense. Most worthwhile opportunities are first encountered and viewed as problem areas. An alert engineer or manager will be able to identify some of these challenges on a daily basis. The real key to success in this endeavor is to broaden the base and increase the input of ideas that can be evaluated as cost-cutting areas in the organization.

Effective cost reduction is a state of mind that must be developed in order to be effective. The employees must feel that management from the top down supports the program. Aside from the support aspect, there must also be the transfer of knowledge that permits employees at all levels to participate in the program. Much of the success that can come from a cost reduction program is derived from target areas of opportunity that are selected for investigation.

TOTAL MANUFACTURING DIVISION

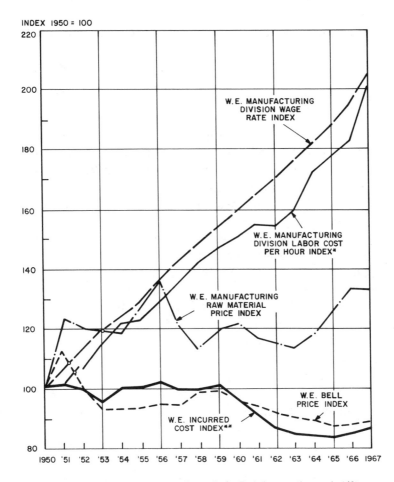

*Average cost per labor hour actually worked. Excludes overtime and shift premiums; includes fringe benefits, incentives, security accruals, etc...

Figure 2-2.   Western Electric has reduced incurred manufacturing costs and bill prices since 1950, even though its wages and raw material prices have risen. (*Source:* Western Electric Company, Inc.)

In order to align our thoughts, we should ask the following typical cost reduction questions.

1. How can labor be utilized, controlled, and measured in a more effective manner?
   • Are labor standards available for use in developing facility re-

quirements and product cost data? If not, can they be utilized?

- Are employees aware of what is expected of them on a daily basis?
- Are employees provided feedback on a regular basis concerning their performance?

2. How can the cost of materials be better managed and controlled?
- Can a lower cost material be utilized? For example, can a metal part be replaced by a molded plastic part?
- Can the dropout or scrap rate at the intermediate and final test position be reduced from 12 percent to 5 percent?
- Can short-interval scheduling be introduced to reduce the material in the pipeline?
- Can ordering, receiving, and storage techniques be improved to reduce inventory investment?

3. Can capital investment be better utilized?
- Can the existing capital investment be measured in terms of utilization? If the answer is yes, is this being done?
- Can large capital requests be deferred or rejected by more effective utilization of existing equipment?
- Can a priority selection system be introduced to insure that the most cost-effective expenditures are being made?

4. Can the use of energy be controlled in a more effective manner?
- Can an energy conservation program be introduced? If one is in use, can it be improved?
- Can periodic audits of energy consumption aid in detection of peak use periods that inflate billing charges?
- Can lighting levels be reduced in the office, shop, and warehouse?
- Can lighting, heat, and cooling be eliminated at nonessential times such as nights, weekends, and holidays?

5. Can expense-related items be controlled in a more effective manner?
- Can items that are now discarded as trash or scrap be recycled in order to reduce or contain costs?
- Can scrap corrugated boxes and certain types or paper be saved and sold for a profit?
- Can service contracts for security, janitorial, and landscape

maintenance be developed around real-need requirements in order to reduce costs?

- Can a more effective selection process in hiring employees be linked to improved training, and reduce the turnover rate in employees?

Take a few minutes to evaluate these suggestions in terms of your job assignment. Record items for further investigations. Once you have done this, the groundwork for a cost reduction investigation has been started. If you have a sizeable list, the next step is to place each idea in a priority list for development.

REFERENCE

1. Criner, E. A., "Cost Reduction—Stepping Stone to Survival in the 1980s," AIIE Proceedings—1980, Spring Annual Conference.

3
Cost Reduction—A Must

The first real signs and concerns associated with an industrial slowdown became apparent in the mid-1960s. It is clear that 1965 was a major turning point in our economy. The late fifties and early sixties produced inflation that averaged 1.5–2 percent annually.

Our position in the 1980s can be described in the following manner: A raging inflation that is without parallel persists at unprecedented peacetime levels. Another challenge is increasing foreign competition in the domestic and international marketplace. Also, at the same time, the United States is burdened with energy-intensive production patterns in a time of rapid upward price movement. The standard of living in the United States is still the goal of most free world nations. Our standard has declined in recent years, while other industrial nations have made positive strides in the past two decades.

INTRODUCTION

The changing conditions and prospects of the U.S. economy are ascribable to many circumstances, but no valid analysis can avoid focus on productivity. This term, much used and much abused, is best understood as the designation for the family of ratios of output quantity to input quantity. In economic parlance, "quantity" is interpreted to include quantities expressed in constant, rather than current, dollars. Such quantities are often characterized as "real." Until now, the productivity subfamily most frequently referred to output per unit of labor input, especially per hour worked or remunerated. It is much easier to measure than output per unit of capital or the more inclusive subfamily of output per unit of labor and capital combined. In the years ahead productivity ratios involving "intermediate" inputs, such as energy and critical materials, must receive greater emphasis. The American Productivity Center has decided to stress the "total-factor" series pioneered by Professor John W. Kendrick.

At least two critical facts stand out when the U.S. position is examined from the productivity standpoint. First, productivity growth has slowed substantially in recent years. This change is reflected in statistics for output per unit of labor and capital combined extending over the past three decades. Figures for output per hour relating to the private business sector show not only a reduction in the growth rate in the 1960s and 1970s but also reversals (negative rates) for 1974 and 1979. The second outstanding fact is that other major industrial nations have exhibited comparative buoyancy. Their productivity rates, as measured by gross domestic product per employed person or by manufacturing output per hour, have grown at robust rates during the same period. Comparatively slow growth rates in labor costs have helped the U.S. trade balance but obviously not enough to offset the adverse effects of petroleum imports.

A strong rate of productivity growth can help counter inflation, provide better cost control, strengthen U.S. trade competitiveness, increase employment, and raise our standard of living. Company officials, workers, and government are currently being challenged to exercise greater self-discipline and to cooperate more diligently in order to achieve better organizational performance with minimum additional costs.

There have been significant differences in the productivity trends for the major sectors of the economy during the past two decades, just as there had been during the earlier years of the century. Similarly, there were significantly varied patterns of sectoral experience in the post-1973 productivity slowdown.

During the entire 30-year span covered by the most recent analysis of total-factor productivity, growth rates for the communications, transportation, agriculture, and manufacturing sectors have been relatively favorable. Construction and mining have lagged, especially since 1973. The trend for public utilities, one of the productivity-growth leaders prior to the OPEC embargo and the energy crisis, slumped badly in the last few years. This was due in part to problems of peak-load scheduling, an absolute decline in demand due to energy conservation, and problems of expenditure for environmental improvement, including conversion of coal to oil and back to coal. Utilities and some manufacturing industry groups were the most severely impacted by forced expenditure of capital for environmental improvement.

This introduction statement by the American Productivity Center cites the problem in terms that are readily understood.[1]

TREND COMMENTS

As you analyze the changes in total-factor productivity by major sectors, 1948–1978, one sector has shown a steady gain. See Figure 3–1. What is responsible for the steady gains that have been made in the communications sector? Also, consider the progress made by the public utilities sector in 1948–1965. During the period 1965–1973 the public utilities had a different or reduced level of achievement. Then the next five years were a disaster as the price of oil continued to rise. The prices of imported oil did not stop in 1978 to coincide with the charted data.

The reverse has been true; oil prices have continued upward. Just to establish a frame of reference, a barrel which cost $2 in 1970 has moved upward to the price range of $33 in first quarter of 1981. We have all seen this impact on the ever-changing utility bills that are re-

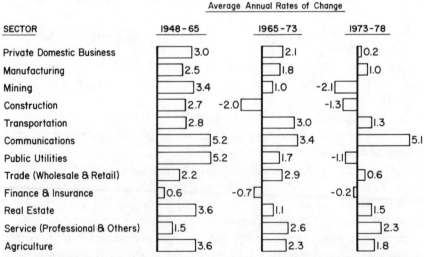

Average Annual Rates of Change

SECTOR	1948–65	1965–73	1973–78
Private Domestic Business	3.0	2.1	0.2
Manufacturing	2.5	1.8	1.0
Mining	3.4	1.0	-2.1
Construction	2.7	-2.0	-1.3
Transportation	2.8	3.0	1.3
Communications	5.2	3.4	5.1
Public Utilities	5.2	1.7	-1.1
Trade (Wholesale & Retail)	2.2	2.9	0.6
Finance & Insurance	0.6	-0.7	-0.2
Real Estate	3.6	1.1	1.5
Service (Professional & Others)	1.5	2.6	2.3
Agriculture	3.6	2.3	1.8

Figure 3–1.   Changes in total-factor productivity for major sectors, 1948–1978. (*Source:* American Productivity Center.)

ceived monthly in the home, business, or manufacturing plant. And today no end is in sight for a leveling-off point.

COST REDUCTION POTENTIAL

An example of things to come in the prices we will pay for gas and electric service is illustrated by a recent filing before the Public Utilities Commission of the State of California. Application Number 60153 was filed on December 23, 1980. It requested authorization among other things to increasing its rates and charges for electric and gas service. The rate requests, if granted to Pacific Gas and Electric (PG&E), would be passed on to all customers as follows:

- Electric rates would increase by $1.138 billion and gas rates by $316.2 million effective January 1, 1982.
- Expressed as a percentage, the electric rates would increase 37.1 percent over the total classes of customers served. Gas rates would increase 9.8 percent. This covers all classes of service including residential, commercial, industrial, and resale.

If industry and residential users survive the first blast, another one is scheduled for January 1, 1983.

- Electric rates would increase by $178.6 million and gas rates by $127.7 million.
- Expressed as a percentage, the electric rates would increase 4.3 percent. Gas rates would increase 3.7 percent. These increases cover all classes of customers served.

Regardless of how the problem is stated, energy will continue to increase in cost in all forms. PG&E would state the problem in another manner. To them, as a seller of energy services, these increases are needed to meet the rising costs of providing service to its customers in this inflationary economy and to enable PG&E to maintain its financial health. PG&E, like any other company, must be financially healthy if it is to be able to attract the capital needed to construct the facilities required to provide electric and gas service to its customers.

A COST REDUCTION APPROACH

PG&E is constantly looking for ways to reduce costs to its customers. Alternate energy sources are being explored to meet demand and provide an economy that may be passed along to its customers.

Geothermal steam, wind, hydroelectricity, solar—PG&E is currently studying and developing these and other forms of alternate energy. Its long leadership in hydro and geothermal development probably gives PG&E more power from alternate energy sources than any other utility.

The leaves you rake each fall could conceivably become part of a renewable organic fuel source called biomass. Biomass is a catchall word for forest and farm residues and aquatic crops that can be burned to make electricity or tapped in other ways for natural gas. PG&E is busy with biomass in a number of different projects.

1. Electricity from farm waste material presents a unique approach. Pellets compressed from grape clippings and other agricultural waste will fuel what is expected to be the world's largest biomass-fueled generating plant. The plant is to be constructed in Madera, about 35 miles northwest of Fresno, California. The project is scheduled to be in operation in early 1982.
2. Gas from steer manure is a novel approach. Manure is being tapped for its methane content at a cattle feed lot in the Imperial Valley. The Department of Energy will provide partial funding to help expand a pilot operation which is run jointly by PG&E and Southern California Gas Company.
3. Gas from garbage offers a unique opportunity for potential savings. PG&E is currently tapping the methane gas formed underground at the Mountain View city dump. Methane is the main and only essential ingredient in natural gas. Methane from perhaps two dozen similar sites in PG&E's service area may provide enough gas for an estimated 100,000 homes.
4. In addition to the three unique applications cited above, PG&E is working with the University of California—Davis to develop further ways to harness the energy potential in biomass.[2]

Changing Sources of Energy

The PG&E Progress, a monthly newsletter, dated February 1981, presented an overview of energy planning to the year 2000. Board

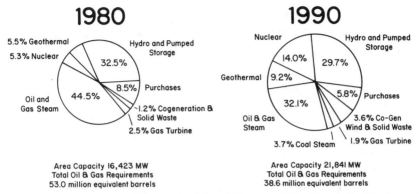

Figure 3-2.   Sources of PG&E area electricity, 1980 versus 1990. (*Source:* Pacific Gas and Electric Co.)

Chairman Frederick W. Mielke, Jr., calls the planning a "base plan . . . a starting point on which the company is basing its long-term plans for an uncertain future," and states that "it will be modified as circumstances indicate the need for change."

The new approach will meet forecast customers' needs but is dependent on improvement in rate regulations and moderation of the inflation rate. It is hoped that this environment will permit PG&E to raise the capital necessary to carry out the plan. See Figure 3-2.

Renewable and alternate resources will play major roles in meeting the ever-changing customer needs. Fewer projects will be required, however, because growth in the use of energy is expected to slow down as a result of increased conservation. PG&E now anticipates electric use growing 2.1 percent a year instead of 3.8 percent predicted if no conservation should occur and gas use growing 1.3 percent yearly instead of 3.3 percent.

Even with fewer projects, the company still expects difficulties in raising the capital to finance these projects. As you may have guessed, the public utilities are capital intensive as well as energy intensive. Alternative and renewable resources such as geothermal, hydroelectricity, wind, and biomass, as well as cogeneration, will play a big part in the year ahead. More than half of the 6.5 million kilowatts of new electric-generating capacity through 1990 will come from such sources.

PG&E plans to continue vigorous development of cogeneration projects, a process whereby the company and its larger customers both use the same fuel source to make electricity and heat or steam to perform useful work for the participating business. A cost reduction case study on energy conservation via cogeneration will be covered in detail in a later chapter.

CHANGING THE DOWNWARD TREND

Cost reduction, a systematic review of the cost inputs, can be the most effective tool available for changing the downward productivity trend. Most companies, both large and small, have experienced a problem in maintaining productivity levels. This has been attributed to many things including the rapid rise in labor cost, energy cost, poor quality, and a host of other related items. Perhaps you are acquainted with these and others.

Effective Cost Reduction

An effective cost reduction program can be one of the most effective tools available for reversing the productivity decline. Let us review the program used by Western Electric Company.

Western Electric is an integral part of the Bell System. Its principal function is to manufacture, to uniform Bell Laboratories standards of design and quality, a wide variety of apparatus and equipment, cable, and wire that go into the Bell telephone plant. Western Electric engineers participate with Bell Laboratories design engineers in developing new and improved products for the Bell System. In addition, Western Electric's manufacturing engineers and the scientists at its Engineering Research Center at Princeton, New Jersey, explore and develop new manufacturing methods and processes.

Western Electric engineers and installs central office switching and transmission equipment in Bell Telephone Company exchanges. It also purchases a large variety of communications products and supplies for the telephone companies. It maintains service centers throughout the United States that include shop facilities for the repair of worn or damaged equipment and warehouses that stock material to meet the daily needs of the telephone companies.

In the period 1956–1967, Western Electric's manufacturing cost reduction program had realized a total of about $246 million in first-year savings. Since savings do not usually end after the first year's operation, cumulative savings have been several times this figure. These

INDEX: 1948 - 100

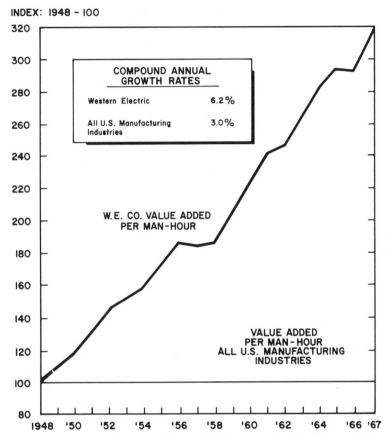

Figure 3-3.   Western Electrics postwar increase in labor output has been substantially greater than the national average. (*Source:* Western Electrics Company, Inc.)

savings have contributed directly to favorable cost- and price-related trends in that time period.[3] These results did not just happen, they were the result of an idea, engineering evaluation, and improved production techniques.

An Accepted Viewpoint

The end objective of any operation that works toward a profit motive is to constantly improve operating results. Profit is the goal and objective of every business enterprise. The achievement of this goal is becoming increasingly difficult. No informed individual will deny that the achievement of profit is linked to increased productivity. This sounds like a very straightforward endeavor, but in reality the link between increased productivity and profit is a very complex relationship. Effective cost control is a must.

Before we plunge into any involvement, a definition of our objective is necessary. Increased productivity is our objective. This objective can be achieved by gaining better control of the items that represent the input equation.

As we achieve more effective cost control we are indeed moving toward our objective. Many managers are riding off in different directions with zeal as they work toward improving productivity. The problems, assumptions, and solutions can become so interwoven that the outcome of the quest for productivity is difficult to evaluate. The job of each manager, engineer, or employee is clearly defined—or if not, it should be—to produce the most product or service with the minimum amount of input.

To achieve maximum output with minimum input should be the goal of any organization. Achievement of this goal is the real measure that defines how effective management has been in a given time period. We should not confuse management potential with proven performance that has been tested and measured over a given time period. Managers in general have the same challenge even though their span of control in a given environment may differ. The daily decision making requires managerial skills, capital, machines, methods, materials, human resources, and last, the organizational climate. Lack of attention to any of these items will produce less than the desired results—as measured by the profit margin.

Today's Environment

In today's environment productivity is linked to efficiency. The most efficiently run companies will usually be the most productive. By concentrating on all operational cost aspects pertaining to the organization it is possible to visualize what can and should be done in the area of cost reduction. Total-factor productivity is one concept that is steadily gaining acceptance. In this concept we are concerned with a given output and the sum of the inputs that were required. As an example, we are concerned with how many units are produced by a given machine, or how many letter pages were typed. Also, a matter of concern is the related cost inputs that were required to achieve a given level of goods or services.

Today's concept of cost control deals with effective cost reduction of the required inputs. It may be more important for one company to control indirect labor, raw material yields, or energy consumption. Another company may have more concern in the area of capital utilization, and in the quality level of the finished product. The key ingredient in either case is the need for effective changes in the operational environment that reduce costs.

HOW TO START LOOKING

Areas of possible cost reduction are not hard to find, evaluate, and change. The key for maximum involvement requires change. We must change our thinking pattern and begin to question all phases of the operation. Cost reduction potential is all around us in every sector of the business environment. The challenge is to select areas for our involvement. Develop a checklist to assist in your search for cost reduction.

The most challenging part of any improvement is not the department or activity to be evaluated, but defining the basic problem. Once the problem is defined, the solutions for solving it can be evaluated. Potential cost reduction can be found in the following areas.

- Standard hours required: Select a sample of typical product lines, examine in detail each segment of the operation.
- Examine material usage on selected product lines. Compare the requirements with actual spot usage.
- Examine the outward-bound quality level. Look for dropouts at the critical processing stages.
- Examine the scrap that is being removed from the operation on a daily basis. Look at the cost of removing the material. Also, look at piece parts that could be recovered for use. Ask how the material ended up as scrap.
- Examine the relationship of direct versus indirect labor in a given product or service.
- Examine the attendance records of all employees in a section, group, or the total operation. Convert the days lost into production dollars. Ask how improvements can be made in this area.

COST:
REDUCTION
POTENTIAL

```
APPLICATION:
    LABOR . . . . . ●
    MATERIALS . . ○
    CAPITAL . . . . ○
    ENERGY . . . . ○
    EXPENSE . . . ○
```

IDEA: Conduct a labor audit on high volume production operations.

APPLICATION: Select a major product line that is currently viewed as an outstanding example - effective cost control, good quality, and meeting shipping schedules. Review specifications, production methods versus manufacturing lay-outs. Utilize standard data, time studies, and random work sampling studies.

SAVINGS: A savings goal of ten percent is viewed as possible in this study. The present group of eleven employees will be reorganized and restructured into a group of ten employees. Estimated annual savings $15,000.

SOURCE: Joint recommendation from the Weekly Quality Circles Meeting.

COST:
REDUCTION
POTENTIAL

```
APPLICATION:
   LABOR . . . . . ○
   MATERIALS . . ●
   CAPITAL . . . . ○
   ENERGY . . . . ○
   EXPENSE . . . ○
```

IDEA: Evaluate the cost of unusable outside-purchased piece parts that do not meet specifications.

APPLICATION: Examine piece parts that are known to cause problems in selected assembly operations. Review incoming inspection plans for acceptance of the piece parts in question. Reject all parts not meeting the incoming inspection levels.

SAVINGS: Improved product sampling will reduce product rework in the assembly operation. Defective material lots can be identified before the invoice is paid. In summary, investment in unusable inventory can be reduced.

SOURCE: Recommendation by the final tester on product line "A."

COST:
REDUCTION
POTENTIAL

APPLICATION:

LABOR . . . . . ◯

MATERIALS . . ◯

CAPITAL . . . . ●

ENERGY . . . . ◯

EXPENSE . . . ◯

IDEA: Conduct utilization studies to eliminate scheduling conflicts on key machines in production operations.

APPLICATION: Develop a list of key machines that are presently bottlenecks in the operation. Examine maintenance records for excessive downtime. Utilize random work sampling to determine - production, set-up time, downtime for maintenance, and other causes.

SAVINGS: Evaluate ways to increase production time. Look for ways to decrease set-up time and maintenance downtime. Improved utilization of key machines can be measured in increased dollars of production per machine hour.

SOURCE: Recommendation from Industrial Engineering Department.

COST:
REDUCTION
POTENTIAL

APPLICATION:

LABOR . . . . . ○

MATERIALS . . ○

CAPITAL . . . . ○

ENERGY . . . . ●

EXPENSE . . . ○

IDEA: Develop lighting standards for the office areas,production areas, and warehouse areas.

APPLICATION: Measure lighting levels in each area. Evaluate energy efficient products to replace the current in-use lighting fixtures. Prepare a schedule to convert to the new lighting standards.

SAVINGS: An expenditure of $108,000 in a large warehouse replaced 900 twin-fixture mercury vapor fixtures. The new installation utilized 900 single-fixture high pressure sodium units. Based on 6,000 operating hours yearly, annual savings of $94,000 were protected

SOURCE: Department of Energy

COST:
REDUCTION
POTENTIAL

```
APPLICATION:
    LABOR . . . . .  ◯
    MATERIALS . .  ◯
    CAPITAL . . . .  ◯
    ENERGY . . . .  ◯
    EXPENSE . . .  ●
```

IDEA: Look for ways to reduce expense in the movement of data from one location to another on a daily basis.

APPLICATION: An electronics company in the Bay Area has a need to move daily progress information from one location to another approximately 30 miles away. A number of surface carriers have been utilized. Estimated daily cost of $50.00 for one trip.

SAVINGS: A comprehensive study evaluated several alternate modes. The final evaluation indicated that carrier pigions with a microfilm capsule attached to the leg could be used. Annual savings of $12,000 per year.

SOURCE: A Bay Area Company

- Examine inventory investment. Look at both work in process and completed goods on the shop floor or in the warehouse.
- Examine equipment utilization. Pinpoint the equipment that may be overloaded or underutilized. Can a different scheduling pattern help alleviate the problem?
- Examine make versus buy decisions that may have an impact on reducing cost.
- Examine machine speeds. Ask what improvements could be achieved by increasing speeds or using a different type of holding fixture for multiple parts.
- Examine the costs incurred in the present distribution network. Look at all aspects of transportation. Evaluate the costs of incoming materials as well as the costs of the completed product.
- Examine the introduction of part-time employees in the organization. Look at placement of part-time employees into groups that have varied work loads, such as accounting, shop assembly, and warehouse.

SUCCEEDING IN TODAY'S ENVIRONMENT

The formula for success in any environment is subject to change from time to time. It is clear that in our present environment the managers and engineers who have the ability to contain cost and improve productivity have the best chance for success.

I would like to share with you a document that has been in circulation since the mid-1950s. This piece of literary achievement is titled "How to Succeed Without Talent." It is doubtful that these rules would be as useful today as they have been in the past. Look back over your career and see if any of these traits have been evidenced by engineers or managers that you have worked with and observed.

HOW TO SUCCEED WITHOUT TALENT

1. Study to look tremendously important.
2. Speak with great assurance; however, stick closely to generally accepted facts.
3. Avoid arguments, but if challenged, fire an irrelevant question at your antagonist and intently polish your glasses while he tries

to answer. As an alternative, hum under your breath and examine your fingernails.

4. Contrive to mingle with important people.
5. Before talking with a man you wish to impress, ferret out his remedies for current problems. Then advocate them staunchly.
6. Listen while others wrangle. Pluck out a platitude and defend it righteously.
7. When asked a question by a subordinate, give him a have-you-lost-your-mind? stare until he glances down, then paraphrase the question back at him.
8. Acquire a capable stooge, but keep him in the background.
9. In offering to perform a service, imply your complete familiarity with the task, then give it to the stooge.
10. Arrange to be the clearing house for all complaints; it encourages the thought that you are in control and helps to keep the stooge in place.
11. Never acknowledge thanks for your attention, this will implant subconscious obligations in the mind of your victim.
12. Carry yourself in a grand manner. Refer to your associates as "some of the boys in our office." Discourage light conversation that might bridge the gap between boss and man.
13. Walk swiftly from place to place as if engrossed in affairs of great moment. Keep your office door closed. Interview by appointment only. Give orders by memoranda. Remember, you are BIG SHOT and you don't give a damn who knows it.

REFERENCES

1. American Productivity Center, "Introduction to Productivity Measurement," Seminar Presentation 1980, Houston.
2. Pacific Gas and Electric Co., "PG&E Progress," February 1981.
3. American Telephone and Telegraph Company, "A Study of Western Electric's Performance," New York, 1969.

4
How To Get Started

This book is primarily concerned with the development, investigation, processing, and introduction of cost reducing ideas into the productivity process. Updating the progress that is being made is essential for informing executives in engineering, production, sales, accounting, and other management areas. The paperwork that conveys this information may vary widely from one industry to another, but this is not important in itself if the system selected works and produces positive results that can be measured and credited to the individual responsible for suggesting the change.

In view of the increased attention that has been paid to cost reduction in recent years, there are still widespread misunderstandings concerning the nature and the meaning of successful cost reduction concepts. The important thing is that the findings of various cost reduction approaches should be assembled in a systematic form and made available to the public in general. Only by increasing our knowledge of the cost reduction approach can we work toward improving productivity and maintaining a competitive position in the United States and world markets.

CURRENT STATUS

The intent of this book is to summarize techniques and procedures concerning cost reduction, with particular references and illustrations that show the workings of an effective program. It is the aim here to show the development, operation, and results reporting that are essential for a successful program. In order to do this we must develop new and improved modes of evaluation.

Applying new technologies, creating, adopting, or adapting work methods, processes, and new concepts in order to obtain increased output for the same input costs must be our goal. Achievement of this goal requires involvement that is creative, imaginative, and well documented in order to protect the originator of the idea. Effective managers structure organizations that generate creative ideas from employees at all levels and involve them in the innovation and implementation associated with technological progress.

In today's economy and the economy of the future, we will survive and prosper only if we, as managers, engineers, and others, can achieve a firm grasp on the major cost inputs in our business. In general terms these inputs are defined as labor, material, energy, capital, and related expense. Efficient control of these key inputs can be traced through the cycle and measured as outputs that can be realized in the final analysis.

Who needs a cost reduction program in today's business environment? The answer is any organization that expects to survive inflation, the quality decline, and the productivity decline that have touched almost every segment of the economy.

COST REDUCTION DEFINED

Cost reduction opportunity will vary with your specific job assignment. Each assignment will present different challenges. However, your assignment as an engineer or manager can be equally rewarding if you maintain an open mind and are always alert to new and innovative ways to reduce costs in your organization.

Let us review some tried and proven areas that will produce cost savings in almost any job you have now or aspire to have at some future date. In broad terms, cost reduction deals with a reduction of labor, material, floor space, energy usage, and expense costs.

- *Labor costs* can be reduced by changing methods, improving material flow, and redesigning tools and equipment.
- *Cost of materials* can be reduced by changing the design of the product, eliminating unnecessary components, and replacing expensive materials with more economical ones.
- *Space requirements* can be reduced by rearranging the work area, using a two-shift operation, and better utilizing "cube" space (floor to ceiling).

- *Energy costs* can be reduced by reducing lighting levels in the office, shop, and warehouse areas; turning off equipment when it is not in use; modifying heating and cooling systems; replacing outdated tools and machines with ones that are more energy efficient.
- *Expense costs* can be reduced by a review of:
  1. Services such as maintenance, security, and janitorial.
  2. The use of expense supplies.
  3. The need for some computer-generated reports that perhaps could be eliminated, combined with other existing reports, or displayed on microfilm and/or microfiche.

In summary, cost reduction is a team effort that involves the company, the employee, and the customer. A successful program has a positive impact on each of the team members: the company, the employee, and the customer.

- *Company benefits* can be found in improved productivity, reduced production costs, improved quality, and increased efficiency. These are the very items that we need to keep pace with outside competition.
- *Employee benefits* for employees involved in cost reduction include personal recognition, job satisfaction, improved salary position, and consideration for promotion.
- *Customer benefits* are where the total impact of cost reduction is evident. The desired end result is in the form of reduced prices without a compromising of quality, and in many instances, reduced lead time to obtain the product.

IDEA INPUT

The input of new ideas is one of the most important elements that can make or break a successful cost reduction program. If you currently work in a cost reduction environment, you are aware of this fact. If you would like to become involved on a daily basis, you must generate new ideas. Also, as an engineer or manager, you must develop a relationship with your peers that leads to an open exchange of new ideas. An atmosphere of trust must prevail in order for others to share ideas with you in an ongoing exchange. A cost reduction audit is a good way to start. See Figure 4-1.

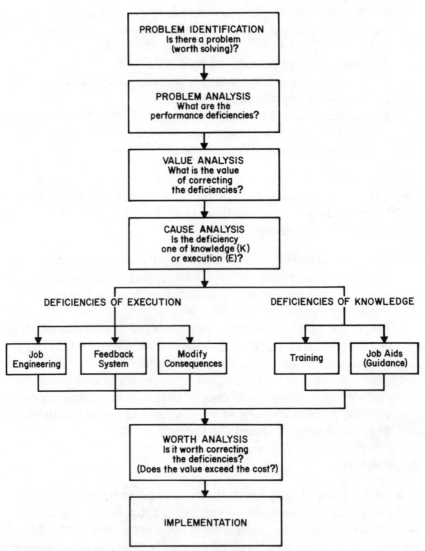

Figure 4–1. The cost reduction audit.

DEVELOPMENT OF IDEAS

Each person may have a different way of developing ideas—from past experience or other learning situations. If you would like to improve your ability to generate new ideas, the following may be of interest.

- Build an idea reservoir. Ideas do not often fall from the blue. You must keep flooding your mind with them by studying related information, by constantly experimenting, and by simulating.
- Carry an idea trap. Carry a pencil and pad with you at all times. A tape recorder may prove to be a good investment for recording your ideas. Why? Ideas are elusive and can quickly depart as readily as they appear.
- Develop a nose for problems. Listen to complaints and gripes from all sources such as the shop, warehouse, and office. Ask the methods department, the long-range planners, or the accountants for comments. Sometimes they may tell you what they think!
- Utilize a checklist approach. Sometimes a structured checklist may be in order. Rearrange? Can we interchange component parts? Could we use another assembly method? Can we use any of the cost reduction techniques from Project X on the new challenge, which is one-half the size and more cost effective?
- Vary your daily routines. Do not fall into the common Monday–Friday rut. Change the way you walk to the shop; walk through the warehouse from a different direction. Observe the number of warehouse trucks that are idle. If your company rents warehouse trucks to supplement their own trucks, perhaps you should study the problem further. As an incentive, consider that one rental warehouse truck equates to approximately $750 monthly or $9000 yearly.
- Be confident, enthusiastic, and open-minded. Your willpower controls your imagination and is affected by your emotions. Build faith in yourself by recording success on selected small ideas before you tackle the big ones.
- Build big ideas from little ones. A well-planned work-order system from one department could be a stepping-stone for plant-wide application. A work measurement approach to reduce expense in a small operation can often be modified and expanded to meet a new challenge.

- Beware of self-satisfaction. Harlow H. Curtice, former president of General Motors Corporation, credited his company's success to the "inquiring mind" approach to problems. "This point of view is never satisfied with things as they are," he said. "We assume that anything and everything—product, process, method, procedure, or human relations—can be improved."
- Learn to spot your mistakes. The person who can spot his or her mistake, find out why it was made, then correct it, is learning how to do things differently. This is how new ideas are generated. Do not be afraid to make a mistake. If you are always concerned about little mistakes, many potential ideas will never be developed.

ORGANIZATION OF YOUR APPROACH

As we approach cost reduction, we must raise certain questions in order to evaluate ideas. In order to get started we need answers to lay a sound foundation.

When a new machine or a process is to be replaced, changed, or added, some logical questions are: Will it do more in a specified time period—per hour, per day? Will it produce a more reliable product? What are the quality versus production trade-offs?

One key question in today's capital-intensive marketplace is, Can it (a new machine, a change in process control) be cost justified? Usually this can be determined early in the evaluation of the idea. In this case we will equate cost justified to achievement to a specific rate of return per dollar invested. As we talk about organizing your approach, let us keep an open mind. Sometimes as we look for approaches that lead to cost reduction, it may be necessary to look for special talent in another department or another organization, and in some cases it may be prudent to seek the expertise of an "outside consultant."

THE CONSULTANT'S ROLE

An outside consultant can prove to be the proper choice for obtaining an overview of your complete operation. Many companies both large and small have used consultants to define specific problem areas. The consultant is usually free to define the problem, propose changes, and define benefits. Also, he or she can remain immune from any political undercurrents that may arise in the course of the investigation. In many cases it may be beneficial to have a local engineering resource, an engineer on your payroll, work closely with the consultant.

The choice is up to the decision maker. This person, or perhaps his or her supervisor, is aware of either a general or a specific problem that needs to be reviewed. Let us say that the decision has been made to conduct a manufacturing facilities overview. The next question to be answered is, Who can be assigned to conduct the study?

Several names from your own engineering group are considered. However, since both of your best-qualified engineers are engaged in a project that cannot be interrupted in order to complete facilities overview, the decision is made to use an outside consultant.

One such outside consultant is Charles H. Olson, P.E., an industrial engineering consultant whose office is in Los Altos, California. He has been active in a wide range of projects for companies in the Silicon Valley, a geographical area in Santa Clara county that is composed of such progressive companies as Advanced Micro Devices, Amdahl Corporation, Apple Corporation, Commodore International, Hewlett-Packard, Intel Corporation, and National Semiconductor. This list covers only a select few of high-technology companies in the Silicon Valley that are known for their ability to produce a quality product at a competitive price.

Let us examine a facilities overview that was prepared by Charles H. Olson. His clients were Xynetics and Electroglas, two companies in the Silicon Valley. The report is in six parts.

1.0 *INTRODUCTION*
   1.1 *OBJECTIVE*
       To evaluate the present manufacturing facility with respect to an overview of material flow, space utilization, material handling, and production equipment. The ultimate objective is to *increase productivity,* that is, to increase the output with respect to the resources input.
   1.2 *METHOD*
       The writer spent two days interviewing managers and supervisors and observing operations. The conclusions presented in this report are primarily qualitative. Each item is presented briefly to provide a base for discussion and additional evaluation.

2.0 *LAYOUT CHANGES*
2.1 *ALTERNATIVE ONE*

1. *PRIMARY MOVES*

    Move the engineering department, document control, and machine shop and gauge room to the Xynetics Building.

2. *SECOND MOVES*

To engineering area:	Production control offices, manufacturing engineering offices, quality assurance offices.
To machine shop:	PCB assembly, precision assembly, 1038, 1024, and 120 assembly, PCB test (depending on proportion of units that flow between assembly and test versus the number that flow between PCB test and systems test), PCB inspection.
To guage room:	Wave solder and PCB cleaning.

3. *THIRD MOVES*

To precision/1038, etc., assembly:	Receiving (starting from overhead door in stockroom), receiving inspection, stockroom receiving.
To PCB assembly and subassembly:	Systems test expansion, subassembly expansion, harness and cable test computer to subassembly area.
To receiving inspection:	Kit audit.

4. *DOCUMENT CONTROL PROBLEM*

    A print file may need to be set up to make available prints-on-demand for manufacturing. This would require a delivery system to service the file, to maintain the supply of prints, add new prints, replace revised prints, and to supply phone requests for special situations. People must be discouraged from going to the Xynetics Building to pick up their own prints. The use of a "terminal digit" filing system will make servicing the print files easier and quicker.

5. *BENEFITS*

a. Increased capacity:	Increases space for all manufacturing areas.
b. Improved flow:	Improves the flow within the subassembly area from subassembly to systems test and from kit audit to subassembly.
c. Increased office space:	Increases space for production control, manufacturing engineering, and accounting.

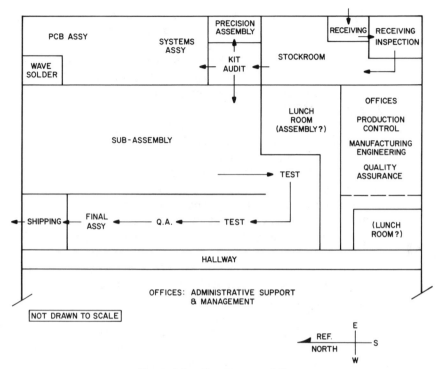

Figure 4–2.   Shop layout and flow.

d. Improved communications:	Improves communications opportunities among the support functions of production control, manufacturing engineering, and quality assurance.
e. Improved environmental conditions:	Isolates the wave solder machine and cleaning from the subassembly area; removes machine shop fumes from the building.

Refer to Figure 4–2.

2.2 ALTERNATIVE TWO

1. PRIMARY MOVES

Move the engineering department, document control, machine shop, gauge room, receiving, receiving inspection, stockroom, and kit audit to the Xynetics Building.

2. SECOND MOVES

To engineering area: Production control offices, manufacturing engineering offices, quality assurance offices.

To machine shop:	PCB assembly, PCB test (depending on the proportion of units that flow between assembly and PCB test versus the number that flow between PCB test and systems test), PCB inspection.
To gauge room:	Wave solder and PCB cleaning.

3. *THIRD MOVES*

To PCB assembly and subassembly:	Systems test expansion, subassembly expansion, harness and cable test computer to subassembly area.

4. *DOCUMENT CONTROL PROBLEM*

A print file may need to be set up to make available prints-on-demand for manufacturing. This would require a delivery system to service the file to maintain the supply of prints, add new prints, replace revised prints, and to supply phone requests for special situations. People must be discouraged from going to the Xynetics Building to pick up their own prints. The use of a "terminal digit" filing system will make servicing the print files easier and quicker.

5. *STOCKROOM SERVICE PROBLEMS*

A "stockroom expediter" may be needed to deliver parts to the manufacturing building to fill shortages or to replace spoilages. The expediter would respond to phone requests. People, expediters in particular, must be discouraged from going to the Xynetics Building to pick up their own parts.

6. *BENEFITS*

a. Increased capacity:	Increases space for all manufacturing areas by at least 50 percent.
b. Improved flow:	Improves the flow within the subassembly area and from subassembly to systems test.
c. Increased office space:	Increases space for production control, manufacturing engineering, and accounting.
d. Improved communications:	Improves communications opportunities among the support functions of production control, manufacturing engineering, and quality assurance.
e. Improved environmental conditions:	Isolates the wave solder machine and cleaning from the subassembly area; removes machine shop fumes from the building.

f. Improved mate- rial control:	Consolidates manufacturing stockrooms for both Electroglas and Xynetics products.

2.3 *LAYOUT PRINCIPLES*

1. As the volume increases, product flow consideration of the layout becomes more important.
2. In a long, narrow room, best use of space usually is obtained by setting benches and test stations so that they run the length of the room.
3. Optimize the layout by providing for the ideal ratios of subassembly to test stations to final assembly work places and maintaining the ratios as the volume of output grows.
4. Provide adequate in-process storage space.
5. Keep aisles clear for free movement of people and products.
6. It is all right to have vacant spaces within the layout. These spaces will be used for future expansion without rearranging existing operations.
7. Provide for storage and control of tooling, assembly fixtures, test equipment, test cables, etc.
8. Proper planning prevents poor performance.

2.4 *MOVE STRATEGY*

1. Develop a *strategic long-range plan* for facilities use.
2. Develop space standards for each production area.
3. Allocate space to each area (production and support) based on the requirements at plant capacity. Prepare a block layout.
4. Develop a strategy for the sequence of moves.
5. Prepare detailed layouts for each area following the sequence of moves.
6. Prepare each area as it becomes vacant. Then move the new occupant into the area over a weekend. This strategy minimizes disruption to production, but it spreads the move and rearrangements over a long period of time.

3.0 *MATERIAL HANDLING*

3.1 *CART SYSTEM*

Use a cart system for transporting kits to production and for moving subassemblies through the production sequence.

1. *SPECIFICATIONS*

   Use the shelf carts with wire shelves. They are lightweight, maneuverable, and easy to maintain. The wire shelves do not collect garbage like solid shelves tend to do. Either the 18″ × 36″ or 24″ × 36″ size would be suitable.

2. *BENEFITS*

The primary benefit is to eliminate handling of parts and assemblies:

Area	Handling Operation Eliminated
Kit audit	Unload incoming kits from stock carts to shelves (by stockroom personnel).
Production	Unload kits from stock carts to staging shelves (by stockroom personnel).
Production	Load carts from staging shelves to move kits to work station (by assembler).
Production	Unload kits to storage shelves at the work station (by assembler).
Production	Unload assemblies to storage shelves at test or final assembly (by assembler or test technician).

3. *HAZARDS*

   a. Providing an inadequate number of carts. Hoarding or hiding of carts may occur. People will spend time looking for empty carts.
   b. Getting empty carts back to the stockroom for loading kits will require time by the stockroom personnel.

3.2 *GRAVITY FLOWRACKS*

Utilize gravity flowracks between receiving and receiving inspection for items requiring inspection.

1. *SPECIFICATIONS*

A single section of gravity flowrack with three or four shelves should be adequate to hold the items for inspection. Space should be allowed for a second section to be added later. The rack is about 8 feet wide by 6 feet deep.

2. *BENEFITS*

   a. Lines up items for first-in, first-out inspection of items.
   b. Keeps work load visible to people in receiving inspection—motivating factor.
   c. Items can be located quickly when expediting is required.
   d. One area of the rack can be designated as the "hot shelf."

3.3 *DRIVERLESS VEHICLES*

Install a wire-guided, driverless vehicle system between the Electroglas Building and the Xynetics Building for transporting parts, materials, and products between the two buildings.

1. *SPECIFICATIONS*

One system is the Electote manufactured by Raymond Corporation. A loading and unloading station would be required at each building. Several trailers can be pulled by the tractor unit.

2. *BENEFITS*
   a. Eliminates a forklift driver to haul materials, parts, and products between the two buildings.
   b. Batches items being hauled between the two buildings.
3. *HAZARDS*
   a. The system is vulnerable to parking lot traffic.
   b. The system is limited to a fixed path.
   c. Items being hauled may be exposed to inclement weather for a longer period of time than with conventional handling equipment.
4. *ALTERNATE APPROACH*
   Use the "elephant train" approach with a forklift for the tractor unit.

3.4 *BUILD QUANTITY*
Establish a "build quantity" as a convenient quantity for counting, issuing, and building assemblies. Use a quantity such as 10 or 20 units.

1. *PROCEDURE*
   a. Bag the build quantity as early in the flow as possible when counted at receiving for parts not being inspected, at receiving inspection, or at stockroom receiving.
   b. Tape or staple bags closed. Ziplock bags may be used.
   c. Label bags. Use peelable labels so bags can be reused, particularly when using ziplock bags.
   d. Stock a few loose parts to replace spoilage, to cover occasional shortages, or to issue for low-volume products when the issue quantity is less than the build quantity.
2. *BENEFITS*
   a. Reduces counting and handling time when stocking, kitting, cycle counting, kit auditing, and assembler is setting up.
   b. Breaks down the production lot size into smaller lots so that the assembler feels he or she has accomplished something more frequently. (How do you get a whale off the beach? Piece by piece.) Provides motivation.

3.5 *KIT SIZE*
Is the present kit size too big?

1. *CONSIDERATIONS*
   a. Is the kit size established for the convenience of production or for the stockroom?
   b. Would a smaller kit issued more frequently smooth out the flow through production?

c. What is the impact of kit size on space utilization?

d. Should some assemblies be decoupled from a "system pull"? PCBs for example?

4.0 STOCKROOM ORGANIZATION

4.1 STOCK ORGANIZATION

1. *PART NUMBER SEQUENCE*

The most common method for organizing stock is by part number sequence. This is the simplest method and the easiest to learn by people in the stockroom. This method generally takes the longest time for order filling.

2. *USAGE*

Parts are classified by usage, A, B, or C. A's are stocked in one area, B's in a second area, and C's in a third area. This method is applicable in a manufacturing environment with a wide variety of products that have a large number of common parts

3. *PRODUCT*

Parts are stocked by product with each product located in a designated area in the stockroom. This method is applicable where there are several products that have few common parts. Order filling time is minimized because the parts are "prekitted" on the stockroom shelves.

4. *ASSEMBLY*

Parts are stocked by assembly with each assembly located in a designated area in the stockroom. Basically, this is a modification of the "product" stocking method. This method is applicable where there are products that have common assemblies. Common parts can be stocked together for assemblies that are similar. Order filling time is minimized because the parts are "prekitted" in the stockroom shelves.

A successful stockroom operation means delivering the right parts to the right place at the right time. This is a corollary of the *success formula:*

Success = Quality Service + On Time + At Least Cost

4.2 STORAGE SYSTEMS

The most common storage system in a stockroom is steel shelving. This system, however, usually results in the lowest utilization of space. Its advantages are that it is least costly, quickest to set up, and quickest to organize.

Other storage systems that utilize space better but cost more:

1. *ROTOBIN*

A rotobin occupies a 3-foot square of space. Each shelf has the equivalent of 10 feet of linear shelving wrapped around in a circle. Rotobins occupy 20–30 percent less floor space than conventional shelving.

2. *MOVABLE AISLE SHELVING SYSTEMS*
Shelf units are set on tracks and are rolled apart to gain access between two units. The units are suitable for storing "C-usage" parts, large parts that do not fit into the main stockroom, packaging materials, and records files.

3. *CAROUSEL STORAGE SYSTEMS*
Carousel systems bring the parts to the person for order filling and bring the shelf to the person for stocking. The system is set up for random access. It is suitable for "A-usage" parts in a high-volume environment. Since it is a computerized and sophisticated system, it is a relatively expensive investment.

4. *ELECTROFILE—VERTICAL*
The electrofile is a mechanical file designed originally for records storage. It is suitable for storing precious metals and small, high-value items because it can be locked. It has a footprint of approximately 8 feet wide by 3 feet deep.

5.0 *AREAS TO UPGRADE*
5.1 *PRINTED CIRCUIT BOARD ASSEMBLY*
Upgrade assembly and processing equipment.

1. *ASSEMBLY BENCHES*
Equipment is available that makes it possible to assemble more than one or two PCBs at a time.

a. Turntable: a custom-built bench with a round top 4 feet in diameter. Eight to ten holders can be mounted on the perimeter. The operator rotates the table as he or she assembles components into the boards.

b. Ferris wheel: a custom-built board holder that rotates vertically. The operator rotates the wheel as he or she assembles components into the boards.

c. Assembly frame: a commercially available fixture that holds a batch of boards for the operator. The bottom row of boards sits in a "U" channel. An "H" channel clamps down on the top edge of the boards and holds the bottom of the second row of boards above the first. A "U" channel across the top holds down the top row of boards.

2. *WAVE SOLDERING*
It was mentioned that the wave soldering machine was inadequate. Considerations include handling component leads and capacity, physically, to handle the larger "mother" boards presently being designed.

3. *AUXILIARY EQUIPMENT*
Equipment is available for preforming and trimming component leads.

5.2 *SUBASSEMBLY AREA*
Design simple holding fixtures for subassemblies.

One operation observed by the writer that appeared awkward to perform was the assembly of harness ends to connectors. A simple assembly fixture could be designed to hold both the harness and the connector to make assembly easier.

Many other methods-improvement opportunities are likely to exist in the subassembly area.

5.3 *XYNETICS TEST*
Enclose Xynetics test to provide an atmosphere free from shop fumes and dust.

A cleaner environment for the test area may reduce test times for both the Xynetics X-Y plotters and the Electroglas X-Y Miniposition plattens and motors. The present area is open to the fabrication and assembly of the X-Y plotters. This is a quality assurance problem that requires more investigation if it has not been done to date.

5.4 *TEST AND QUALITY ASSURANCE*
Systems test and quality assurance appear to be bottleneck areas.

The area was laid out on the principle: "Keep the area small so that the product has to keep moving through." To the writer this is "suboptimization" that says in other words: "To increase product flow through an area, create bottlenecks."

The area needs to be optimized:
  a. Provide space for growth.
  b. Provide space for technicians and engineers to work around the systems.
  c. Provide space for storage of test fixtures, cables, subassemblies, etc.
  d. Provide space for burn-in.
  e. Motivate technicians to follow housekeeping practices.

6.0 *PRODUCTIVITY MEASUREMENT*
6.1 *BENEFITS*
When productivity and performance measurements are introduced into an operation for the first time on a regular, formal basis, a performance increase of at least 20 percent can be expected within six months.

This phenomenon occurs in nearly 100 percent of the organizations when performance measurements are implemented. It is this "secret" that many major consulting firms use to guarantee improvements of from 10 percent to 25 percent and on which they justify some big fees.

The impact is that manufacturing capacity can be increased by 20 percent using the present staff and facility, everything else remaining constant.

Productivity and performance measurements can be implemented using production and payroll information available. Engineered time standards do not need to be developed but may be a refinement to the measurement system one or two years hence. An effective system can be set up using the historical information presently available.

Definitions:

$$\text{Productivity} = \frac{\text{Output Units}}{\text{Resources Input}}$$

$$\text{Performance} = \frac{\text{Target Time to Produce Output}}{\text{Actual Time}}$$

6.2 *A SUMMARY BY THE AMERICAN PRODUCTIVITY CENTER*

The measurement of productivity is essential to both understanding and improving productivity. Through measurement, an organization can determine its level of productivity, analyze its strengths and weaknesses in terms of productivity performance, and evaluate trends and progress toward improving productivity.

Despite this importance, techniques for measuring productivity are generally poor and widely misunderstood, particularly at the company and plant levels. Labor productivity statistics are published regularly by the Bureau of Labor Statistics (BLS) for various industries and sectors of the U.S. economy in the form of "output per employee-hour" or "output per employee." The Department of Commerce publishes related microeconomic statistics such as GNP per capita. No measures, however, that include all factors of production are published by the government, and it has not recently computed company or plant productivity data. Several private studies, published irregularly, have been the only efforts made in this area.

"Total productivity" is defined as a measure of the efficiency with which the quantities of all factors of production (inputs) are utilized to produce the quantities of an organization's total products and services (outputs). Therefore, it is the ratio of total outputs divided by total inputs. A family of "partial productivity" measures can also be developed by establishing ratios of total outputs to one or more input categories, such as output/labor, output/materials, and output/capital. A third type of productivity measure is "value-added productivity" in which the output less raw materials is related to part or all of the other inputs. These definitions can be expressed in equation form as:

$$\text{Total Productivity} = \frac{\text{Total Output}}{\text{Total Input}}$$

$$= \frac{\text{Total Output}}{\text{Labor} + \text{Materials} + \text{Energy} + \text{Capital}}$$

$$\text{Partial Productivities} = \frac{\text{Output,}}{\text{Labor}} \quad \frac{\text{Output}}{\text{Materials}}, \quad \frac{\text{Output,}}{\text{Capital}} \text{ etc.}$$

$$\text{Value-Added Productivity} = \frac{\text{Total Output} - \text{Raw Materials}}{\text{Labor} + \text{Materials} + \text{Energy} + \text{Capital}}$$

The above definitions are stated intentionally in terms of inputs and outputs as opposed to financial purchases and sales. Productivity is a measure of the efficiency of the process by which factors of production are converted into final products and services. Consequently, raw-material and finished-goods inventory changes and financial transactions that take place outside the conversion process are not included in the productivity calculations.

All productivity measures are ratios of output quantities to input quantities. Some partial productivity ratios can easily be stated in quantitative terms: tons per man-hour or units per BTU. However, in other partials and in total productivity, unlike quantities must be combined: tons and gallons of products; employee-hours, pounds, BTUs, etc., of inputs. This problem is resolved by using a set of weights, representative of the relative importance of the various items, to combine unlike quantities. Base-period prices are the recommended weights to be used for total productivity calculations, although other weighting systems such as "man-hour equivalents" are frequently used. By summing of the current quantities multiplied by their respective weights, and dividing by the sum of the base-period quantities multiplied by the same weights, the relative change in quantities from the base period to the current period is determined. This calculation is performed separately for the outputs and the inputs. The output/input ratio of these results is the relative change in productivity from the base of the current period.

When base-period prices are used as the weighting system, the product of the current quantity times the base price yields the deflated value (in base-period dollars) of the current quantity. In situations where relevant quantity and unit price data are not available (such as capital costs), the same results can be obtained by deflating the current value by an appropriate price index. While individual company or plant price indexes are preferable, the detailed price indexes published by the BLS can also be used for this purpose.

It is assumed in price weighting quantities that the relative prices reflect relative quality. Whenever price does not properly reflect quality (such as today's controlled energy prices) or when quality changes occur between the base and current periods, appropriate quality adjustments must be made to avoid distorting the productivity ratio.

A model for measuring total and partial productivities, based on the above principles, has been developed by the American Productivity Center. In its present form, the model is primarily applicable to company and plant level measurement for manufacturing industries where the outputs are physical products. Productivity measurement in service industries and in the administrative and support departments of manufacturing businesses involves the difficult problem of defining and quantifying the service output. Also, in addition to efficiency measurement, the importance of output effectiveness (e.g., of the services performed) must be more actively considered for service and administrative functions.

The very process of measuring and analyzing productivity results creates an important awareness and concern for company productivity improvement, which in turn favorably affects the ultimate objective: bottom-line profit improvement. Each company must determine for its particular business the type and extent of productivity measurement which provides the necessary degree of detail and accuracy to achieve productivity improvement results. Regardless which measurement techniques are determined best for a particular situation, productivity measurement is an important and necessary element of a comprehensive productivity improvement program.

PROBLEM IDENTIFICATION

An effective cost reduction program is built on a foundation of problem identification, a clear statement of the problem, a proposed statement of change, and selling of the idea to management. See Figure 4-3.

Is there a problem (worth solving)?

How do you know you have a problem?

☐ What do you observe that indicates there is a problem? Labor costs appear high, shop efficiency is down, production bottlenecks, scrap rate is rising, perhaps a warehouse flow problem?

How will you know when the problem is solved?

☐ How will things look different... reduced cost, improved quality?

☐ What numbers will increase or decrease... rejects, increased production?

What are the "stakes" involved?

☐ How will the solution of this problem affect the quantity, quality, customer satisfaction, cost of the product or service being offered?

☐ What would be the economic impact of making the lowest 25% of the performers as effective as the top 25%?

How long has this been a problem... short term or long range?

How general a problem is it?

☐ Where does it occur... shop or warehouse?

☐ When does it occur... random or sequence?

☐ How frequently does it occur... daily or weekly?

☐ Does it not occur in all locations... if so why?

Figure 4-3.   Problem identification.

After a problem has been identified you must decide if it is a problem worth solving. Certain signs may be present; they can include abnormal labor cost, a drop in shop efficiency, production bottlenecks, a rise in scrap rates, or a warehouse flow problem. This is one of the most interesting stages in the development of a cost reduction case, problem definition.

You may be surprised to learn that the toughest person to sell your ideas to is yourself. To hit a home run, or conclude a successful cost reduction, you first must get the idea past the ever-present mental blocks that we all have. Unless you act on your ideas, nothing is going to happen. Time and again the person who succeeds is not the one who thinks up the best ideas but the one who can act on the ideas available.

Consider the facilities overview that dealt with material flow, space utilization, material handling, and production equipment. Make a mental list of some possible cost reductions that could be developed.

5
Organizing For Cost Reduction

Small businesses outnumber large ones in every country of the free world. Therefore it is important to be aware that the application of cost reduction techniques can be applied in both large and small organizations. The results of a successful program can be measured in any business or service environment, although a smaller organization can usually be equated to a simpler set of operating guidelines.

In general, a successful cost reduction program will entail the following elements:

1. Management involvement—this is essential for all employees to know that management is open to change.
2. A defined organization—the leaders in the cost reduction effort need to be identified with the roles they perform.
3. A definition of cost reduction—a definition of what is and is not to be considered "fair game" in the quest for cost reduction.
4. Goal setting—employees need to be aware that goals have been set. One organization may choose to establish as a goal a reduction of $100,000 in overhead expense. Another organization may establish 4 percent of forecasted production as a goal.
5. A measurement system—most cost reduction systems will be concerned with first-year savings and also a projection for five years. In some industry segments the five-year data may not be relevant.
6. Employee participation—this is a key element in any ongoing program. Management must develop new and innovative ways to focus interest on the program.
7. Program evaluation—this can be accomplished in a number of different ways. A monthly status report can be a timely feedback device.

The seven elements of organization can be designed to suit the needs of any size company or organization.

THE PUZZLE

An ongoing cost reduction program is essential to improving productivity. Increased productivity can usually be linked to profitability in any type of business enterprise. Such a program seeks to reduce overall product cost. This can be achieved by a systematic evaluation of requirements that impact on the end product. Significant cost reductions can be achieved through commonsense applications in daily operations. Also, major unique innovations will lead to large savings when identified, developed, and installed. See Figure 5-1.

- Define clearly the objectives and responsibilities of cost reduction. This can usually be accomplished in a written instruction for use at one or more company locations.
- Design forms and procedures that meet the needs of your operation. This can include forms that document the basic idea and idea development, and a summary form that includes the final action taken.
- Utilize techniques such as value analysis, work simplification, and cost avoidance; use the approach that fits your needs. Record

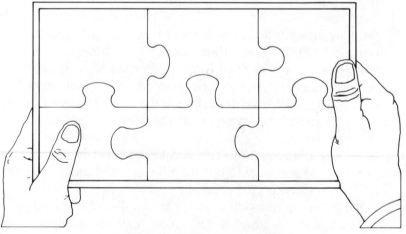

Figure 5-1.   The productivity puzzle.

your progress daily. Good notes are essential in conducting a cost reduction case. Cover each action you take in such a manner that it can be used as documentation at a future date.

- Develop the habit of looking for improvements. Ask: Is it being done the best way? Is it necessary? See Figure 5-2.

DATE 12/16/80

I. PRESENT METHOD

CONTRACT WAREHOUSE SPACE IS UTILIZED IN A COMMERCIAL FACILITY TO STORE CERTAIN TYPES OF SHORT AND LONG TERM MATERIAL. CHARGES ARE BASED ON A ONE TIME IN AND OUT FEE PLUS A MONTHLY CHARGE BASED ON WEIGHT. SPECIAL TRUCKING IS REQUIRED FOR PICK UP AND DELIVERY.

---

2. I PROPOSE THAT

A COMPLETE INVENTORY BE TAKEN TO DETERMINE IF CERTAIN ITEMS CAN BE JUNKED. ALSO, THAT THE ITEMS REMAINING TO BE STORED SHOULD BE STORED IN LEASED SPACE. THIS SPACE SHOULD BE NEAR THE SERVICE CENTER SO THAT RECALL TIME CAN BE FAST AND TRANSPORT CHARGES MINIMAL.

---

3. IT IS ESTIMATED THAT SAVINGS WILL BE REALIZED IN THE FOLLOWING AREAS:

A MONTHLY BILLING CHARGES FOR STORED MATERIAL
B. REDUCED TRANSPORTATION CHARGES
C.

SIGNED E. A. CRINER   DEPT. 820130   TEL. X4496

ADDRESS REGIONAL CENTER

WITNESSED & UNDERSTOOD BY   R. P. MECK

ATTACHMENTS: ___ YES   X NO

SHEET I OF 1 SHEETS

Figure 5-2.   Cost reduction proposal.

CASE NO. 411,079
ISSUE NO. 1
PAGE 1 OF 3 PAGES

TITLE (BRIEF DESCRIPTION)

REDUCE WAREHOUSE STORAGE COST BY CHANGING
STORAGE MODE.

KEY WORD INDEX
1. REDUCE
2. WHSE. STORAGE
3. COST

SCOPE AND OBJECT

THIS CASE DEALS WITH THE REMOVAL OF MATERIAL
NOW STORED AT THE CHIPMAN WAREHOUSE IN
SAN LEANDRO. THE MATERIAL WILL BE
INCORPORATED INTO THE PROPOSED PLUG-IN
POOL CONSOLIDATION IN SAN LEANDRO

LOCATION PACIFIC REGION (NCNSC)
ORGANIZATION NO. 22PC820130
PROBABLE COMPLETION DATA 9/81
CLASSIFICATION
☒ COST REDUCTION   ☐ DEFINITE ROUTINE
☐ OTHER DEVELOPMENT   ☐ GENERAL ROUTINE

ASSOC. PLANT AUTH. NO. _____
DATE APPVD. BY C.R. COMM. _____
RATE OF RETURN _____
MAKE VS BUY APPROVAL _____

Figure 5–3. Investigation case authorization.

- Secure management endorsements on all cost reduction efforts. This can usually be done with a brief discussion between you and your supervisor. A memo should be written to cover the who, what, where, when, how, and why. This is essential before the formal paperwork and approvals are secured. See Figure 5-3.

THINK COST REDUCTION

Increased productivity through cost reduction has become headline news. Leaders from government as well as buisness have cited the poor performance of the United States in productivity improvement in the last decade. However, reading between the lines, it is clear that many of these leaders are not too sure exactly what they are deploring. Productivity is a widely misunderstood concept. It is related to virtually every other economic concept in one way or another, but most of the standard economic solutions do not directly address productivity.

Improvement in productivity through cost reduction can mean increased profits, higher overall wages, more jobs, and a generally higher standard of living. Productivity improvements benefit all segments of our economy: labor, management, consumers, and government. However, there are many problems associated with productivity that are effectively blocking progress. Our basic growth rate is slowing down as the population growth changes, foreign competitors are making inroads resulting from lower labor costs, our production costs are increasing due to continual inflation, government regulations are increasing in areas such as safety, and there is a general lack of understanding about productivity.

Remember that we are considering cost reduction in light of comparing what it costs the old way with what it would cost to introduce the new way. See Figures 5-4 and 5-5.

411,079

DEVELOPMENT EXPENSE:

    ENGINEER'S SALARY: $4,100

ESTIMATED COST AND SAVINGS:

A. PRESENT OPERATION

    THE CHIPMAN WAREHOUSE IN SAN LEANDRO IS USED FOR OVERFLOW
    MATERIAL THAT CANNOT BE CONTAINED IN THE PRESENT WAREHOUSE
    AT THE NCNSC.

    BILLING TO THE SERVICE CENTER IS BASED ON A WEIGHT BASIS OF
    MATERIAL STORED AT CHIPMAN. WEIGHT OF MATERIAL STORED IN 1980
    VARIED FROM A HIGH OF APPROXIMATELY 8.0 MILLION POUNDS, AND
    A LOW OF 3.5 MILLION POUNDS.

    CHARGES ARE BASED ON A ONE TIME CHARGE OF $1.06 PER 100 POUNDS
    FOR AN IN-OUT CHARGE, PLUS A CHARGE OF $.42 PER 100 FOR THE
    REMAINING MONTHS. THE OTHER COST FACTOR IS TRANSPORTATION
    CHARGES TO AND FROM THE CHIPMAN WAREHOUSE.

    ESTIMATED CHARGES OF THE PRESENT METHOD ARE ESTIMATED AS FOLLOWS:
        STORAGE CHARGES   - $289,000
        EST. TRANSPORTATION -   35,000
                    $324,000

B. PROPOSED METHOD

    AN ESTIMATED NUMBER OF SELECTED PALLETS (3365) COULD BE REMOVED
    FROM THE CHIPMAN WAREHOUSE. THESE PALLETS CAN BE STORED IN THE
    BACK OF THE CONSOLIDATED PLUG-IN POOL OPERATION IN APPROXI-
    MATELY 30,000 SQ. FT.

Figure 5-4. Development factors.

411,079

B. PROPOSED METHOD: (CONT'D)

   A) 3365 PALLETS X TURN OVER 4 TIMES PER YEAR
      IN-OUT CHARGE OF $1.06 PER PALLET/100 LBS. AND STORAGE CHARGE
      OF $.42/100 LBS. PER MONTH.

      3365 X ($1.06 X 700 LB) + #($.42 X 700)
      3365 X (7.42 + $8.82) = $54,647

   B) TURNOVER 4 TIMES PER YEAR

      $218,588 = $54,647 X 4 TIME/YR

      ITEMS (A & B) ARE AN ENGINEERING COST ESTIMATE OF THE MATERIAL
      BEING STORED AT THE CHIPMAN WAREHOUSE.

   c) COST ESTIMATE TO STORE THE SAME MATERIAL AT THE NEW PLUG-IN
      POOL IN SAN LEANDRO (YEARLY ESTIMATE)
      1) SPACE COST 30,000 SQ. FT. AT $3.00 = $ 90,000
      2) TRUCK RENTAL $750 MO.   =   9,000
      3) LABOR - WAREHOUSE PERSON   =   25,000
                                               $124,000

   D) SAVINGS 1ST YEAR
      $94,588 = $218,488 - $124,000
                   (B)   (c)

FIVE YEAR SAVINGS PROJECTION:
YEAR 1 $ 94,588
YEAR 2   102,155
YEAR 3   110,327
YEAR 4   119,153
YEAR 5   128,685
           $554,908 ÷ 5 = $110,981
YEAR 2 - 5 BASED ON AN 8% INCREASE PER YEAR

NOTE   TRANSPORTATION CHARGES ARE ASSUMED TO BE THE SAME AS EXISTING.

Figure 5-5. Methods versus savings.

One way of making this comparison is to consider two factors: (1) the cost of putting the new method into effect and (2) the annual savings that would be realized if the new method were already in effect. If these two factors are known, a simple evaluation can be made of whether the idea is worthwhile.

Costs for putting an idea into effect can be divided into major categories as follows (see Figure 5-6):

1. *Development expense:* Development expense includes the investigating engineer's salary for time spent in the development

ESTIMATED COST AND SAVINGS DATA

I. DEVELOPMENT EXPENSE
    (A) ENGINEER'S SALARY LOADED .... $ 4,100 ....
    (B) EXPERIMENTAL SHOP WORK 0
    (C) EXPERIMENTAL PLANT OR
        DEVELOPMENT FACILITIES 0
                                     TOTAL .... $ 4,100

2. ASSOCIATED EXP. TO BE INCURRED
    AS A RESULT OF THIS CASE
    (A) MOVES AND REARRANGEMENTS 0
    (B) REMODELING PLANT 0
    (C) EXPENSE TOOLS AND SUPPLIES .... $ 2,000
    (D) ENGRG. SERVICES 0
    (E) SERVICES, OTHER THAN ENGRG. 9,000
                                     TOTAL $11,000

3. NEW PLANT REQUIRED AS A RESULT OF
    THIS CASE (EXCLUDING ITEM 1(C))
    (A) LAND, BLDGS., LAND IMPRV.,
        MACH. & TRANSP. EQUIP.
        (INCL. DESIGN) 0
    (B) SMALL TOOLS 0
    (C) FURN. AND FIXT. $160,000
                                     TOTAL $160,000

4. TOTAL EXPENDITURES (ITEMS 1, 2 & 3) $175,100

5. ANNUAL SAVINGS BASED ON
    (A) 5 YEAR AVERAGE $110,981
    (B) CURRENT LEVEL* .... $ 94,588
    COST ESTIMATE NO. _____1_____ DATE 12/80
    APPROVED BY: _____ ORG. NO. _____

    * 12 MONTHS SUBSEQUENT TO THE DATE SAVINGS ARE MADE EFFECTIVE

	LAND. BLDGS., LAND IMPRV. MACH. & TRANSP. EQUIP.	SMALL TOOLS FURN. & FIXT.
6. PLANT TO BE REPLACED		
(A) COST		
(B) COST OF DISPOSAL		
(C) SALVAGE VALUE		
APPROVALS: PRELIMINARY		
FINAL _____ DATE _____		

Figure 5-6.   Estimated cost and savings data.

of the case. It also includes the building or acquiring of experimental facilities for trying out or determining the feasibility of the proposed method. Every case will require expenditures in this category.

2. *Associated expense:* This may include expense tools, supplies, and services other than engineering. It also might include the changing or modification of existing facilities. Some cost reduction cases will require expenditures in this area.

3. *New plant (facilities):* This is the cost of new land, buildings, machinery, furniture, and small tools. A smaller number of cases you may be involved with will require funds in this category.

The objective of any cost reduction program is to produce savings by a change in methods, procedures, or material, or a physical change in design of the end product. You may find other applications in the work assignment you deal with daily. The name of the game is develop an idea that solves a problem, define the changes that are to be made, make the changes, complete the documentation, and close the case with a specified amount of savings. Usually savings are expressed as annual savings—for the first year through the fifth year. Also, it may be desirable to state a five-year average which may include a factor for inflation.

A PROGRAM REVIEW

Beech Aircraft Corporation of Wichita, Kansas, has always been interested in reducing costs. They had learned through previous experiences that there are no miracle solutions to the total productivity problem. However, the year 1975 was a turning point in their cost reduction/productivity improvement program. It has often been said that necessity is the mother of invention. No doubt you have heard this many times before.

Real progress in their quest for cost reduction can be traced to contract negotiations between the union and the aircraft company. It was in these meetings that the company shared their concerns regarding declining productivity with the union. As in many related discussions between the union and the company, the union was quick to point out that it appeared that ideas from the shop floor were not really wanted.

After the dust had settled both parties were in agreement that a joint endorsement of the cost reduction program was essential for a successful program.

Program guidelines were designed in 1975 and are still in use today. The program is structured on the concept that input ideas from each employee are required in order to keep the program moving forward.

- All new employees are provided a one-hour orientation about the program. The new employees can bring fresh ideas to the job. The orientation provides insight on how to handle processing of the idea.
- A full-time productivity council staff, with a manager, administers the program.
- A resource panel member is available in each division of the company to answer questions and provide guidance.
- Timeliness is recognized as a very important ingredient. Prompt feedback on ideas is provided.
- The program is open to all employees in all departments and all job classifications.
- Joint support of management and labor is essential to the successful operation of the program.

The Flow of Ideas

To be effective the program must process new ideas on a daily basis. When an idea is being developed an employee does not have to look for help. Productivity council representatives are always available to assist. The representatives are elected by their fellow workers for a one-year term. This is in addition to the employees' normal job assignment. As a general rule there is one representative for approximately 30 employees.

The representative is available to assist every employee in developing the idea, filling out the required paperwork, and processing the idea. After the idea has been presented in the initial form, a council evaluator meets with the employee and leaves a typed copy of the paperwork involved. This transaction provides the employee with something tangible. It also reflects that the company is interested in listening to new ideas.

Within one week after the idea is submitted, the proposal is reviewed by the council representative, council evaluator, and departmental supervisor. In the evaluation three questions are asked.

- Is the idea practical?
- Can the idea contribute to increased productivity?
- Does the idea have a 50-percent chance of adoption and use?

All proposals that pass the initial screening entitle the suggestor to a $10 early recognition award. This action informs the employee that the idea has been reviewed and has been judged to have merit. About 60 percent of all ideas submitted receive the $10 award. Feedback is provided on all ideas even if they cannot be implemented. Items that are evaluated and placed into operation are eligible for additional awards. Additional payments up to $100 can be authorized by the council's recognition committee. All awards are reviewed twice a year for employees who have previously earned $60–$100. Additional awards ranging from $100 to $1000 may be awarded based on the final idea evaluation.

Recognition extends beyond the cash awards. Other forms of saying thanks include informal dinners for council representatives.

Is the Program a Success?

If the input of ideas is a measure of success, the program is a success. The productivity council has presented over $1.0 million in awards over a 5½-year period. The acceptance rate of ideas is approximately 40 percent. A total of 21,800 ideas have been accepted and implemented. Bottom-line dollar savings are a measure of success. There are other items that must also be evaluated—better communication between labor and management, improvement in morale, increased job satisfaction. While they cannot be measured directly, they are important in the overall picture.[1]

ATTRIBUTES OF A SUCCESSFUL PROGRAM

A well-defined cost reduction program will have the following objectives:

1. To stimulate and coordinate cost reduction and development in the organization.
2. To provide a medium for exchange of new ideas and methods.

3. To minimize duplication of cost reduction and development work.
4. To provide recognition by management for specific accomplishments.
5. To evaluate the effectiveness of cost reduction effort.

A successful cost reduction program is a means of publicizing an idea and giving official recognition to those who exert the effort to bring their ideas to fruition. Often an idea is hard to formulate. At times it is hard to express in words. Sometimes it is based on untested assumptions. The cost reduction program is designed to help you overcome these and other difficulties. The first precept of the program is, "Get your idea on paper." To do this a universal form has been furnished to help you clarify your idea. It is programmed to help you answer these questions (see Figure 5–2):

1. What is your idea all about? This is simply to identify in other peoples' minds the general nature of what you are talking about, whether it is a new routine or the modification of an existing one.
2. What is the scope of your idea? Is it a change that applies to only one specific problem? In other words, how big is your idea?
3. What specifically is your objective? What are you trying to do? How do you expect it to be accomplished?
4. What steps do you feel will have to be taken to see it accomplished?
5. How long will it take to do it? See Figure 5–7.
6. What is your best estimate of the costs involved to accomplish it? Materials? Services of others?
7. How much do you expect to save? How worthwhile is your idea in terms of dollars and cents? Some ideas, you say, cannot be evaluated in advance. Agreed! But whenever possible, estimate expected savings.

In summary, a cost reduction program provides:

1. A means of crystallizing an idea into a plan of action.
2. A means of evaluating the economic impact of an idea.
3. A means of obtaining management sanction for expenditure of time and money in developing an idea.

OUTLINE OF WORK AND SCHEDULE

OPEN CASE	1/81
LAYOUT APPROVAL	1/81
ORDER PALLET RACKS	2/81
RECEIVE PALLET RACKS	5/81
INSTALL PALLET RACKS	6/81
TRANSFER MATERIAL	7/81
REVIEW FINAL SAVINGS	8/81
CLOSE CASE	9/81

LOCAL DIVISION OF SAVINGS

100% TO NCNSC

NO APPLICATION AT SCSC & WIOSC

SUGGESTED BY	CONDUCTED BY
E. A. CRINER - SV	E. A. CRINER - SV D. R. SEAMANS- NCNSC

REASON FOR REISSUE

Figure 5-7. Outline of work and schedule.

4. A means of exchanging ideas between individuals, departments, organizations, and other divisions within a company.
5. A means of providing recognition to individual initiative and creativity of the engineer or manager.

COST ASPECTS OF THE PROGRAM

Your first reaction may be that such a system sounds good. This may be followed by a question: Can we really afford to become involved in such an operation? This may be an oversimplification of the problem. The real question may be, Can your company continue to function without such a program?

The cost reduction program is the systematic study and development of each worthy idea aimed at reducing these costs. The suggested change that is proposed must be innovative and cannot simply be the result of a change in company policy, or a reduction in business volume.

Today's environment, with high inflation and interest rates, higher material and labor costs, higher taxes, greater costs due to government regulations and controls, and increased competition from sources that manufacture items for the electronics industry, dictates that we continue to make significant improvements in our cost reduction results.

HOW TO START YOUR PROGRAM

Organizing for the venture into a cost reduction program is dependent upon the scope, type, and size of the company involved. One engineer may be the entire effort devoted to cost reduction in a small company. The approach to problems selected for evaluation, action on the problems, and record keeping can be a very simple endeavor. Results obtained in this situation can be very rewarding for the individual and the company.

The charter in a small organization could be defined in one simple sentence, "Come up with ideas, recommendations, and suggestions on how to reduce costs." If you were the only engineer in this environment this could be a rewarding experience. However, this approach is very limited in application. Most engineers and managers are tied to a predetermined interface, since most are employed by larger companies and in many cases cover a restricted job assignment. In some cases the entire job assignment may be directed to a component part of an end product.

The small company or organization can conduct an effective program with a simple set of procedures. Teamwork is still a key requirement in any type of organizational structure. An engineer in a small company will, by the nature of the resources available, need to be more self-sufficient. A greater overall knowledge will be required since the exchange of information and ideas may be limited. This, in fact, is not all bad and can provide a wide area of challenge and growth on the job.

More direct contact with management can usually be expected in the small company. This is to be expected due to the reduced number of management levels. In general, you will have more exposure to the final decision maker, the individual who will either accept or reject your cost reduction proposals. Your chances of success in this mode of operation may be improved due to the direct input from management in the idea stage and problem definition stage. To illustrate how this may work, consider the following entry into a typical case.

The plant manager has a staff meeting every Monday to review the production results from the previous week. One item of interest has been a small custom-printed circuit board. The board is produced for an outside vendor. For the past six months production has been normal in relation to meeting schedules, cost control, and quality. For the past two weeks, schedules have slipped, quality has declined to below normal, key components have been on back order. This is unusual since the labor force has been stable, supervision has not changed, and production schedules have been held constant. An initial evaluation by management has established this as a priority item. As the meeting is concluded the decision is made to conduct an engineering evaluation. Your assignment is to complete an initial review for next week's meeting.

Your first reaction may be to start grasping for a wide variety of factors associated with the problem. However, this could be time-consuming and could distract from the real improvement and problem definition that could be achieved in the specified time period. You may be ahead if you use this approach:

1. Analyze the present operation from start to finish. Look for bottlenecks and trouble spots. A flowchart may be helpful in locating problems.

2. Break down the job in terms of operations, inspection, transports, and storage. Record the total distance traveled. List this as the present mode. Did you note anything that should be changed?
3. Question and classify each detail. Ask the questions: What? Why? When? How? Where? Who? Did you note items that could lead to the present problem?
4. List your observations. Did you see anything that could be changed?
5. Prepare a brief but factual report for management of your findings and recommendations in order to alleviate the present problem.

The chances are that your investigation will lead to one or more cost reduction ideas that can be developed. Keep telling yourself "there is always a better way." Write up the details into a formal case. Discuss the improvements proposed with all concerned. Secure the necessary approvals to proceed. Keep going until the new or better way is developed.

A CONTRASTING APPROACH

The charter for a larger company that has multiplant operations, a well-organized distribution network, and a decentralized engineering department still has the same basic objectives. In general the objectives may be stated as follows:

- To stimulate and coordinate cost reduction development in all segments of the operation.
- To exchange new ideas and methods among divisions in the company.
- To reduce duplication of cost reduction and development work.
- To provide a maximum return for each cost reduction dollar expended.
- To provide individual recognition by management for cost reduction accomplishments.

In a larger organization the processing of a cost reduction proposal will follow a more formal routine. As an example, meetings for processing ideas or opening cost reduction cases are usually scheduled on a monthly basis. In order for your case to be on the agenda for discussion, it must be reviewed by a number of individuals in advance of the meeting. The normal flow may include the department coordinator, the department chief, the subbranch coordinator, and the cost reduction committee chairman.

This approach may appear to take away from the freedom that is available to the engineer in a smaller company environment. However, this can be more than offset by the support and expertise of other experienced engineers who can share information with you. Once your cost reduction has been approved you are free to proceed in the evaluation, using your experience and the experiences of other engineers who may be able to share their knowledge with you.

FOLLOW YOUR PROGRESS

Once you have the idea, it should be documented and witnessed on the "proposal worksheet." Any other related material such as drawings, sketches, or photographs should be attached. The proposal should define the present method. See Figure 5-8. The plastic parts mentioned here are a wide variety of telephone set handles and housings that are assembled in the Western Electric repair centers. The plastic parts are refinished using an automatic paint process with a number of outside suppliers. A careful study of the material flow indicated that this small process change could have substantial savings.

After an initial evaluation the formal cost reduction case was opened. See Figure 5-9. The formal case contains the title, scope and object, and other related information dealing with the formal opening of the case with the cost reduction committee. The presentation is usually made by the engineer who suggested the idea. This is the time when members of the cost reduction committee can ask questions. Who should be better able to answer any questions that may arise than the engineer or manager who did the initial work on the case opening?

The scope of work and the time schedule are covered next. In general, this is a brief activity and time schedule. The activity defined and the schedule are usually left to the engineer. However, there may be times when the investigation runs either ahead or behind the published schedule. See Figure 5-10. Your progress will be dependent on your application, and the involvement of others who may be working with you. My experience has been that successful cost reduction can-

DATE 6/6/79

I. PRESENT METHOD

PLASTIC TELEPHONE PARTS ARE PAINTED BY AN OUTSIDE SUPPLIER. EACH
PART IS WRAPPED SEPARATELY IN A SPECIAL PAPER TO PROTECT THE
SURFACE FINISH. THE PARTS ARE THEN INSERTED INTO SEPARATORS IN
THE SHIPPING CARTON.

---

2. I PROPOSE THAT

ALL PLASTIC PARTS, HANDLES, AND HOUSINGS BE SHIPPED FROM THE
SUPPLIER ON CUSTOM FORMED STACK TRAYS. TRAYS WOULD BE SUPPLIED
BY WESTERN.

---

3. IT IS ESTIMATED THAT SAVINGS WILL BE REALIZED IN THE
FOLLOWING AREAS:

A. A REDUCTION IN PAINT CHARGES PER UNIT
B. A SAVINGS IN UNPACKING AT ASSEMBLY LINE
C. A SAVINGS IN INVENTORY CONTROL

SIGNED E. A. CRINER   DEPT. 820130   TEL. X4496

ADDRESS REGIONAL CENTER

WITNESSED & UNDERSTOOD BY   E. L. MAHLER

ATTACHMENTS: ____ YES   _X_ NO

SHEET I OF 1 SHEETS

Figure 5-8.   Another cost reduction proposal.

not be done by one individual working in a vacuum or in a dark room.
Success is usually enhanced with the input of others. The input need
not come from one who is an expert in a given field. Sometimes an
open mind may surpass a wealth of special knowledge in a narrow
spectrum.

CASE NO. _411,005_

ISSUE NO. ___1___

PAGE _1_ OF _1_ PAGES

TITLE (BRIEF DESCRIPTION)

STACK TRAYS FOR USE WITH HOUSINGS AND
HANDLES

KEY WORD INDEX

1. _STACK TRAYS_

2. _TRANSPORTATION_

3. _STORAGE_

SCOPE AND OBJECT

THE SCOPE OF THIS CASE DEALS WITH THE MAXIMUM UTILIZATION
OF PLASTIC STACK TRAYS FOR THE TRANSPORTATION AND
STORAGE OF REFINISHED TELEPHONE PLASTIC PARTS.

THE OBJECT IS TO ACHIEVE A REDUCTION IN THE COST OF
REFINISHING PLASTIC PARTS DUE TO SUPPLIER SAVINGS IN
LABOR AND PACKING MATERIAL. ADDITIONAL LABOR SAVINGS
ARE REALIZED FROM THE REDUCTION OF EFFORT REQUIRED TO
UNPACK THE PARTS FOR PRODUCTION USAGE.

LOCATION _____ SOUTHERN CALIFORNIA _____

ORGANIZATION NO. 22PC820130

PROBABLE COMPLETION DATE _10/81_

CLASSIFICATION

[X] COST REDUCTION   [ ] DEFINITE ROUTINE

[ ] OTHER DEVELOPMENT   [ ] GENERAL ROUTINE

ASSOC. PLANT AUTH. NO. _____

DATE APPVD. BY C.R. COMM. _____

RATE OF RETURN _____

MAKE VS BUY APPROVAL _____

Figure 5-9.   Another investigation case authorization.

Last, but not to be overlooked, is the estimated cost and savings that are projected for the case. Each organization will have a different labor rate for engineering effort. This rate times the number of hours you estimate on the case is recorded in the block on the right. See Figure 5-11. Any other expenses should be recorded in the appropriate location.

A combined total of expense is achieved by adding the various subgroups. Savings are estimated on a first-year as well as a five-year basis. The rate of return, which can be described as a go-no gauge, will have a different meaning at different points of time. The basis for this change will be the cost of capital and the tax rate for the specific company. Approaches for developing a rate of return will be covered in a later chapter. The rate of return in a specific case can lead to an accelerated or revised schedule as you become involved with several investigations at the same time.

The lack of capital funds in any organization, large or small, will have an effect on the performance of a successful program. Marginal cases may have to be deferred until a later date. These decisions can be made by your cost reduction committee. In a market dealing with rising money cost, a grading system for each case presented may be required. Since each industry is different, you will need to develop guidelines to fit your needs.

```
┌─────────────────────────────────────────────────────────┐
│  OUTLINE OF WORK AND SCHEDULE                             │
│  REQUEST CONTRACT SPECIFICATIONS,                         │
│  ESTIMATE POTENTIAL SAVINGS              11/79            │
│                                                           │
│  PRELIMINARY INVESTIGATION               12/79            │
│                                                           │
│  DETERMINE TRAY REQUIREMENTS AND                          │
│  ORDER TRAYS                              5/80            │
│                                                           │
│  COMPLETE INVESTIGATION                   6/80            │
│                                                           │
│  VERIFY FINAL DATA AND PREPARE CLOSING                    │
│  REPORT                                   7/80            │
│                                                           │
│  TAKE PARTIAL SAVINGS FOR HOURSING        8/80            │
│                                                           │
│  FINAL CLOSING FOR PLASTIC HANDLES        6/81            │
│                                                           │
│  LOCAL DIVISION OF SAVINGS:                               │
│                                                           │
│  100% TO WIOSC                                            │
│  NCNSC HAS A SEPARATE CASE     (411,005-144)             │
│  SCSC HAS A SEPARATE CASE      (411,005-145)             │
├───────────────────────────┬─────────────────────────────┤
│  SUGGESTED BY             │  CONDUCTED BY                 │
│                           │                               │
│  E. A. CRINER - SUNNYVALE │  E. A. CRINER - SUNNYVALE     │
├───────────────────────────┴─────────────────────────────┤
│  REASON FOR REISSUE                                       │
│                                                           │
│                                                           │
│                                                           │
│                                                           │
└─────────────────────────────────────────────────────────┘
```

Figure 5-10.   Another outline of work and schedule.

ESTIMATED COST AND SAVINGS DATA

1. DEVELOPMENT EXPENSE
   (A) ENGINEER'S SALARY LOADED — $ 2,100
   (B) EXPERIMENTAL SHOP WORK — 0
   (C) EXPERIMENTAL PLANT OR DEVELOPMENT FACILITIES — 0
   TOTAL — $ 2,100

2. ASSOCIATED EXP. TO BE INCURRED AS A RESULT OF THIS CASE
   (A) MOVES AND REARRANGEMENTS — 0
   (B) REMODELING PLANT — 0
   (C) EXPENSE TOOLS AND SUPPLIES — $ 9,675
   (D) ENGRG. SERVICES — 0
   (E) SERVICES, OTHER THAN ENGRG. — 0
   TOTAL — $ 9,675

3. NEW PLANT REQUIRED AS A RESULT OF THIS CASE (EXCLUDING ITEM I (C))
   (A) LAND, BLDGS., LAND IMPRV., MACH. & TRANSP. EQUIP. (INCL. DESIGN) — 0
   (B) SMALL TOOLS — 0
   (C) FURN. AND FIXT. — 0
   TOTAL — 0

4. TOTAL EXPENDITURES (ITEMS I, 2 & 3) — $11,775

5. ANNUAL SAVINGS BASED ON
   (A) 5 YEAR AVERAGE — $35,370
   (B) CURRENT LEVEL* — $35,370
   COST ESTIMATE NO. ___1___ DATE ___10/79___
   APPROVED BY: _____ ORG. NO. _____
   * 12 MONTHS SUBSEQUENT TO THE DATE SAVINGS ARE MADE EFFECTIVE

	LAND. BLDGS., LAND IMPRV. MACH. & TRANSP EQUIP.	SMALL TOOLS FURN. & FIXT.
6. PLANT TO BE REPLACED		
(A) COST		
(B) COST OF DISPOSAL		
(C) SALVAGE VALUE		
APPROVALS: PRELIMINARY		
FINAL _____ DATE _____		

Figure 5-11. Another estimated cost and savings data.

PROBLEMS AND CHALLENGES

Each type of industry is different and somewhat unique. However, each industry tends to present challenges that are the same yet very different in the final analysis. See Figure 5-12. A cross section of business in general is illustrated here. Your exposure to problems in various areas such as the coal industry, sawmills, steel, synthetic

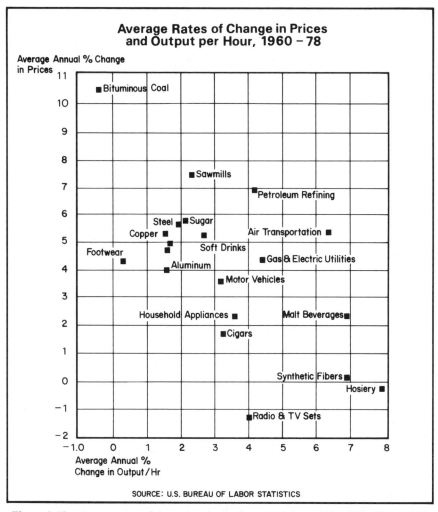

Figure 5-12.   Average rates of change in prices and output per hour, 1960–1978. (*Source:* U.S. Bureau of Labor Statistics. Courtesy American Productivity Center.)

fibers, motor vehicles, or hosiery will be quite different from one another. One thing is certain, though: You will identify problems. These problems can be broken down into the basic elements that relate to productivity. Look for unique applications that deal with labor, material, energy, capital, and expense. Each of these items can be used as a stepping-stone leading to increased productivity.

Your assignment in any industry segment, while it is unique, can also be very similar to that of engineers and managers in a wide range of other industries. Our common goal is to provide a product or service, with high quality, and at a competitive price.

Just take a minute and list five ideas that you feel can be developed into cost reductions within the next year. Is this too optimistic? If so, let us settle for three at this point.

Now, let's put those three on paper!

REFERENCE

1. Hanssen, G.M., "Productivity Gains From the Shop Floor Up," *Production Engineering,* March 1982.

6
Setting Target Goals

Productivity improvement through effective cost reduction is essential to the survival of our free enterprise system. It is the best tool we have to fight inflation, reduce unemployment, increase profits, provide capital, and improve the quality of working life in America. It has application in all companies, both large and small. How do you achieve success in a cost reduction program? This is certainly a valid question if you are involved in an existing program or would desire to form a new program. Let us examine some proven techniques that have been used.

MAKING WAVES

The best advice that I personally ever received concerning success in a cost reduction program came from one of my first supervisors. He said "Don't be afraid to make waves. . . . You should do this for the sake of the organization." People should make waves, they should not be afraid, timid, or shy about this kind of challenge. If you are convinced you are right, you have a right to try it out. In fact, you have an obligation to try it out for yourself and the company. In summary, just because the "boss" or "upper management" does not understand the proposal, do not give up.

You might have to repackage the data in a format that can be grasped by someone who is removed from the problem that you are trying to define. Effective communication is very important in defining new concepts associated with cost reduction involvement. Keep it simple yet informative. Keep this fact in mind: Upper management is usually far removed from the problem areas that can generally be defined as true cost reduction. However, their interest in the program

JUNE 30, 1981

TO ALL SUPERVISORS

IN REVIEWING OUR COST REDUCTION PERFORMANCE THUS FAR THIS
YEAR, IT APPEARS DOUBTFUL THAT WE WILL MEET OUR OBJECTIVE,
UNLESS A GREAT DEAL MORE ATTENTION IS PAID TO THE GENERA-
TION OF COST REDUCTION IDEAS. I AM CONFIDENT THAT IF EACH
OF YOU TAKE AN OBJECTIVE LOOK AT YOUR OWN OPERATIONS, YOU
CAN FIND COST REDUCTION OPPORTUNITIES.

THIS YEAR, TO DATE, WE HAVE PAID OUT OVER $15,000 IN SUG-
GESTION AWARDS. THE LARGEST OF THE AWARDS RESULTED FROM
SUGGESTIONS THAT WERE SIMPLE AND YET INNOVATIVE. IT'S
DIFFICULT TO UNDERSTAND WHY THE SUPERVISORS HAD NOT SEEN
THE OPPORTUNITIES A LONG TIME AGO.

I AM REQUESTING THAT EACH SUPERVISOR COME UP WITH A COST
REDUCTION PROPOSAL TO REDUCE LABOR AND ONE TO REDUCE MATERIAL
USAGE OR JUNKING, WHERE THESE OPERATIONS ARE UNDER CONTROL OF
THE SUPERVISOR. I WOULD LIKE TO HAVE THIS COMPLETED BY
JULY 22, AND WILL REQUEST A REPORT FROM EACH SUPERVISOR ON
THEIR FORECAST OF POTENTIAL IDEAS.

MANAGER

Figure 6-1.   An invitation to submit your idea.

can usually be captured by a well-developed proposal. You might
think at this point that management is removed from the role of generating ideas—this is not true.

Sometimes management does take the initiative to get the attention
of supervisors and engineers. This can be achieved through an invitation to submit your ideas for consideration. See Figure 6-1. This may
not be required in a small organization in which management is visible
on a daily basis. However, in a large organization this mode of communication can be very effective.

COST REDUCTION INVOLVEMENT

Cost reduction is everybody's business. It is an absolute must. It is a
way of life that we must adopt regardless of the type of business, size
of company, or geographic location.

How do you do it? Is it simply a matter of working long hours?
Showing initiative in certain key areas of the shop or warehouse
operation? Perhaps the key is to become skilled in the evaluation of
expense operations. Some people would say it is just being lucky—

being at the right place at the right time. Do not count on the aspect of blind luck too much. Cost reduction involvement is like most other involvements. The output can usually be related to the input effort that you expend. What is the key to a successful program?

The answer is: all of the above—and more. Probably the most important ingredient is setting reaching goals and knowing how to work well with people. Working with people includes your peers and others with whom you interface daily. Most cost reduction applications cut across numerous departments in typical organizations. For an engineer or manager, the typical organizations dealt with may be production, accounting, marketing, engineering, quality control, long-range planning, and in some cases, a corporate headquarters group.

THE GOAL-SETTING CYCLE

You will find from time to time that natural opportunities for planning and goal setting occur:

- The beginning of a new year provides an excellent opportunity to establish reaching goals.
- The quarterly review of achievement provides a natural time to review and update goals.
- A new project or major change in your job assignment provides a natural time for adjustment of goals.

In addition to the three natural goal-setting times for long-range involvement, your short-range goals will be influenced by the following: the need to reduce costs in problem areas, the need to reorganize operations, the need to improve the quality of output, and the need to increase your own performance level.

A certain amount of planning is needed to cope with any of these short-range challenges. Keep one thing in mind! What is not planned in January will not be accomplished when December rolls around. In summary, long-range planning coupled with the establishment of reaching goals is a must in the quest for cost reduction.

THE DESIRED RESULTS

How can this planning and goal setting be translated into cost reduction results? This is indeed a valid question. Let us examine some typical examples of cost reduction. These results did not just evolve. They are the result of problem definition, planning, and follow-through for a wide variety of problems in a cross section of today's business enterprise.

These examples may be far removed from your field of involvement. Regardless of this, one thing is for certain. Your desire to proceed in a program to contain costs is the first step in the achievement of positive results.

An Example of Labor Savings

Examine your packing and shipping areas for potential savings. You may be surprised at what is really going on in that segment of the operation.

Carlson Craft, a printing firm in North Mankato, Minnesota, did examine their packing and shipping area. The company produces a wide variety of thermographic and offset printed products for distribution to more than 20,000 dealers in the 50 states plus Canada and Puerto Rico. A cross section of products includes business cards, personalized announcements, stationery, and entertainment items.

Management found it necessary to develop more productive methods of shipping the finished product. An analysis in one area indicated that sealing the corrugated cartons was time consuming and expensive. The old method of sealing corrugated cartons required seven employees to manually seal outgoing orders. A gum parchment tape that was water activated was being used. In some cases large cartons had to be cut down by hand in order to accommodate the shipment.

After the new random case sealer was installed, two people were processing between 3000 and 3500 shipments on a daily basis. The result was a savings of five people. Based on the cost factors associated with the changeover in equipment versus labor savings, the payback period was approximately four months.[1]

A worthwhile cost reduction? Yes, indeed.

Energy Savings—Unique Applications

The rising cost of jet fuel has provided a major challenge for all airlines. Picture this, for instance: The kerosene that fueled jet engines in 1973 cost about $.11 per gallon. In 1981 the same fuel exceed $1.00 per gallon. This single expense item had become a major consideration. How can a 1-percent reduction in fuel be translated into savings?

On an industry basis airline spokesmen indicate that a 1-percent fuel reduction can be equated to yearly savings between $86 million and $100 million for all major U.S. air carriers. This potential is a ray of sunshine in an era when unprecedented losses in the range of $400 million occurred in 1980. The airlines have introduced a number of unique ideas that are planned to have a direct impact on this critical budget item that has increased from about 10 percent to 50 percent in the short period of eight years.

Some of the new conservation techniques would have been hard to implement before the cost crunch. As in other areas of involvement timing is very critical. Now the time is right to consider ideas that would have been discarded in the past. A cross section of the new ideas are:

- Eastern Airlines has removed a mechanical stairway from its Boeing 727 fleet. This move reduces 410 pounds of weight. This could be the equivalent of three paying passengers.
- United Airlines has removed an old-style movie screen and installed a lightweight model that works just as well. The new screens are 27 pounds lighter.
- TWA removed its Boeing 707s from trans-Atlantic crossings and assigned them to domestic routes. An employee noted that the movie projectors had not been removed. Removal of two projectors from each aircraft amounted to approximately 180 pounds on each plane.
- United Airlines has discontinued filling the 800-pound tanks that provide water for the lavatories on 747 aircraft. This change in procedure caused no real problem since the unused water supply had been dumped at the end of each flight.
- American Airlines recently converted to new lightweight fiberglass buffets. The result of this change can be credited with saving 2.5 million gallons of fuel per year. A related program, a conversion to lightweight seals, is credited to another savings of 1.0 million gallons of fuel.
- TWA had long maintained a policy of having cleaning crews place a copy of the TWA Ambassador magazine in each seat

pocket, even if the previous passenger had not removed one. As a result, as many as two or three magazines often collected in the seat pockets. In some cases, the extra magazines could amount to as much as 500 pounds. This procedure has been changed and now only one copy per seat pocket is allowed. Savings in fuel and the cost of the publication have been substantial.

The old saying is still true today—"Time is money." Looking for the most economical cruise speed is very important today. An extension of a few minutes in flight time is more cost effective in terms of fuel consumption. In summary, each extra pound of weight consumes about 29 gallons of fuel per year.

A simple change that substituted emergency slide-raft combinations saved 2200 pounds per airplane. For a fleet of 18 Boeing 747s, the savings amounted to $710,000 yearly for TWA.[2]

Controlling Expense Operations

When a company can hold down operating costs, it can control price increases. These benefits can be passed along to the customers. Amway is constantly searching for such cost-effective measures in order to streamline warehouse operations. One such item recently installed is called TILTS: Total Inventory Location and Tracking System. With the introduction of the TILTS program, Amway no longer has to take the time-consuming and expensive annual inventory.

Keeping an accurate account of the location and quantity of each item in an inventory is a major undertaking for even a small company. At Amway they are currently tracking over 10,000 item descriptions. The new system works due to employee training, employee attitudes, system design, and management support. Results from 150 worldwide locations with 98.9 percent accuracy are coupled with over one million input–output transactions a month.

The daily cycle is completed in the following manner:

- An employee records what is in storage against a computer balance that has been generated by the TILTS printout.
- Information from each warehouse is keyed into a diskette, then converted to magnetic tape for computer storage.

- The diskette reader transfers raw TILTS data to the computer.
- Updated inventory reports are generated for the next business day.

The system, by eliminating the annual inventory, has generated substantial cost reduction savings in the range of $500,000 yearly.[3]

Cost Reduction Results

A successful cost reduction program is characterized by the input of new ideas. The flow of ideas into a cost reduction program can originate from many varied sources. Let us review several programs with which you may be familiar:

- Employee suggestion system: This system has worked well in Bank of America, Intel, Shelmac, Vari-Tronics, Western Electric, and a wide spectrum of other companies in manufacturing, banking, and retailing, and even on the college campus.

The paperwork to document the idea may vary widely or, in some cases, the idea may be handwritten in a simple note to the individual designated as the system coordinator. The purpose is to state a present method of operation, define a better procedure, and document an action plan.

- Results can vary. Handling checks is nothing new for the Bank of America. An assistant vice-president was awarded $50,000 for his idea suggesting a way to streamline the bank's check-processing procedures and thus to speed collection of checks drawn on other banks. Checks were being held at a Bank of America center in Pasadena until the film was developed so they would be available for refilming if there was a problem with the negatives. The procedure was changed to release the checks after filming. This simple idea reduced the bank's daily "float" of uncollected checks by $4.8 million. The result is an annual pretax savings of $703,000. BOA's employee suggestion system pays up to 10 percent of expected first-year savings to a $50,000 maximum. In this case, an assistant vice-president of the bank's data center was the first employee to win the top prize.

The award was made by the chairman of the bank's executive committee. The event was publicized in the *San Francisco Sunday Examiner and Chronicle* on May 10, 1981.[4]

• Western Electric's Employee Suggestion Program generated record rewards of almost $7 million in total savings in 1980. This figure is more than 60 percent higher than the 1979 level. Nearly 20,000 suggestions were generated. More than 5000 ideas were adopted. Nearly $900,000 in awards were divided among the suggesters.

The largest savings from the program were generated at the Omaha Works, with first-year savings exceeding $1.3 million from the 368 adopted suggestions. One suggestion from a toolmaker resulted in a $500,000 savings and a $10,000 award. The input ideas that have been proposed range from simple changes to very complicated ones. [5]

• Central Methodist College, located in Fayette, Missouri, a small school of approximately 750 students, recently introduced a cost savings incentive program. The program is open to faculty, students, and staff employees. Refer to Figure 6-2 for an example of how to stimulate idea input.

The program is designed to make everyone more aware of ways to help the college save money through reduced overhead expense and energy conservation. Suggestions are submitted to the president's office for review. After a determination that the suggestion would be beneficial, an award of 10 percent of the projected savings is made. Response to the newly installed program has been good. [6]

WANTED

1

Good Idea

Figure 6-2.   Wanted—one good idea.

ENGINEERING COST REDUCTION

Engineering cost reduction is defined as the systematic study and development by engineering personnel of each worthy idea. Emphasis in the last few years has been directed at reducing the cost of production operations and material in manufacturing. Warehousing activities have also offered unique challenges for technological advances. The real challenge in today's environment is how to define, measure, and evaluate the ever-growing body of clerical, technical, and management personnel that has increased rapidly in most companies.

CONTINUED INVOLVEMENT

What is the reason for cost reduction? The answer to this question can be found in almost any industrial publication. At least one article in each publication will be devoted to cost reduction/productivity improvement. This emphasis on cost reduction is usually brought about by the problem of meeting the demands of three groups:

- The buyer requires quality, service, and a competitive price.
- The worker demands wages commensurate with the change in the cost of living and community wage standards.
- The stockholder expects a reasonable return on investment.

In facing these problems a company must choose between lower profits or a reduction in manufacturing costs. In granting an increase in wages, the company must accept a decline in earnings or an increase in the selling price, or reduce the cost of manufacturing the product. Regardless of the end result, quality, service, and competitive price are the three factors that are still required regardless of the manufacturer's decision. Therefore, it is necessary to search constantly for cost reduction opportunities to meet these demands.

In seeking opportunities for cost reduction, consideration must be given to the division of responsibility for costs of customer's orders, land, building, machinery, materials, and labor. Our efficiency, and consequently our effectiveness, is dependent upon our ability to bring these elements together with the least possible time utilization and the greatest possible customer satisfaction.

Consequently the cost involved in each of the manufacturing elements must be a joint responsibility of labor, engineering, and management. One could go through each department and each function and show that an effective cost reduction program involves a team effort with everyone working toward a common goal. The industrial engineering function has become an important member of this team in most companies. If your company's size cannot support a full-time industrial engineer, consider using a consultant to develop your cost reduction plan.[7]

Your personal efficiency can be improved by planning assignments in an orderly manner. Refer to Figure 6-3. A work sheet such as this could aid you on a daily basis. The sheets can be filed for 30, 60, or 90 days to measure your progress. I have used this technique for a number of years and found it very helpful. If this does not appeal to you, the daily diary may suit your needs. Regardless of what you use,

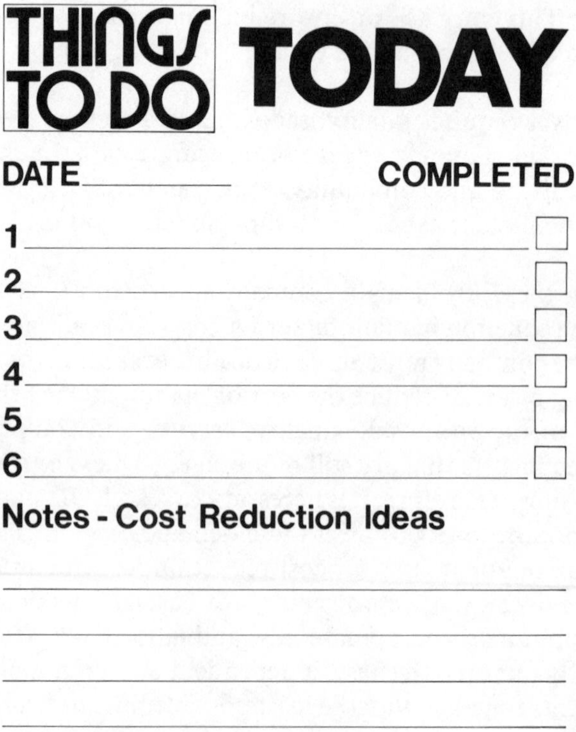

THINGS TO DO TODAY

DATE_____   COMPLETED

1_____ ☐
2_____ ☐
3_____ ☐
4_____ ☐
5_____ ☐
6_____ ☐

Notes - Cost Reduction Ideas

_____

_____

_____

_____

Figure 6-3.   Things to do today.

it is imperative that ideas and concepts for future investigation be recorded for reference.

COST REDUCTION CONSIDERATIONS

Examine your daily work environment. Look for ways to improve ideas that are in your area of responsibility. Record some of the items that may have potential for an in-depth review. Typical examples that you may develop may be included in the following list:

- Look for ways to improve manufacturing operations. Ask if operations can be combined. Can electrical and mechanical tests be combined?
- Investigate the purchase and use of less expensive materials.
- Examine the flow of material through the assembly operation, inspection, packing, and into the shipping area.
- Study the present layout in shop, warehouse, and office area and develop a more efficient one.
- Evaluate computer operations and determine the impact on production operations. Would a real-time production control be beneficial? Can it be cost justified?

FEEDBACK TO MANAGEMENT

Effective feedback to management is an essential part of a successful cost reduction program. Information flow should be informative, concise, and easily read by management. The president of a clothing company has two rules that deal with the engineers in his company:

- Rule 1: He never reads any memo or proposal over one page long.
- Rule 2: He never listens to any presentation over 20 minutes in length.

As a member of management he is not unique. Most managers are busy, and your approach will affect your success in communicating with them. The sample trip report in Figure 6–4 may be useful. It can be modified to fit your use.

TRIP REPORT

DATE: _____   NAME: _____

LOCATION VISITED: _____

PERSON OR PERSONS VISITED: _____
_____

OBJECTIVES: _____
_____
_____
_____

RESULTS: _____
_____
_____
_____

RECOMMENDATIONS: _____
_____
_____
_____

COST REDUCTION POTENTIAL: _____
_____
_____

Figure 6–4.   Trip report sample.

Why Use the Trip Report?

There are several valid reasons for using the trip report. The report is brief yet covers all of the essentials that are needed to make a positive statement. The statement can cover an inquiry, define a problem that you are working on, or document a project that will be developed at some future date. This format has been helpful to me in the pursuit of cost reduction and on consulting assignments.

The report can be handwritten in order to save time and also the expense of a secretary. It provides excellent documentation for reference at a later date.

CASE STUDY EXAMPLE 1

Case Description	Productivity/Cost Reduction
Company	General Foods Corporation
Data Source	American Productivity Center 123 North Post Oak Lane Houston, Texas 77024
Estimated Savings	See Individual Tables

An ongoing cost reduction program is a tool that can be utilized in the development of a productivity measurement system. The inputs such as labor, material, energy, capital, and other related expense can be isolated and reviewed for cost reduction potential. Each segment of industry will have unique features that require special considerations. Your involvement on a daily basis will dictate which inputs offer the best potential for cost reduction.

This case study furnished by the American Productivity Center illustrates in a graphic manner the relationship between cost reduction and improved productivity.

Plantwide Productivity Measurement at General Foods Corporation

What do Super Sugar Crisp, Maxwell House Coffee, Open Pit Barbecue Sauce, Minute Rice, Jello Pudding, Gravy Train Dog Food, and Tang Beverage Mix all have in common? Yes, they are all edible and, yes, they are all manufactured and distributed by General Foods Corporation (GF), a $5 billion plus multinational corporation that operates around the world to provide food products of all types for practically any occasion. But another, perhaps more interesting common feature of these food items is that they are all subject to the same yardstick for measuring the efficiency by which they are produced. They are all covered by General Foods Plantwide Productivity Measurement Program.

I. Program Development

In 1975 the corporate management addressed itself to the question: Can we measure on a total-plant basis how effectively a plant uses the resources made available to it in the form of materials, labor, and capital in turning out production? If we can, how should we do it?

A group composed of cost accountants, industrial engineers, and manufacturing operations employees was formed to address this basic question. The result of their investigation was a Plantwide Productivity Measurement Program (PPMP).

The principal feature of the PPMP is the computation of a Productivity Index (PI) for each plant. The PI for a year is the ratio of the productivity during that year and the productivity during a base, that is, the current-year productivity divided by the base-year productivity. It is an indicator of the plant's productivity improvement from year to year. Tested in 1975 on two large plants (one capital intensive and the other labor intensive), the PI is presently being computed at 20 of the 34 major plants at GF, with more plants being added each year.

GF employs close to 50,000 men and women at more than 100 locations in 16 different countries. Every shopping day consumers purchase some 23 million packages of GF products. In producing these products for the marketplace, GF has for many years maintained an objective of being the low-cost producer in the food industry. The PI has become a valuable indicator of the company's progress toward that goal.

II. Description of Productivity Index

The PI for an individual plant portrays the overall management effectiveness at the plant. Since productivity is defined as output divided by input, the productivity index is the ratio of output and input with both factors adjusted for inflation. The output (the numerator of the productivity calculation) is conceptually the cost value that should have been added by converting raw and packaging materials into finished goods if plant operations were carried out at base-year efficiency levels. It is, therefore, the value (in base dollars) added to the raw materials that are put into the production process.

The input (denominator of the productivity calculation) is the amount of effort and resources expressed in base-year dollars used to achieve the output. The input elements that combine to form the total input amount include:

- Direct labor.
- Service labor—maintenance, quality control, etc.
- Administrative and clerical manpower.
- Purchased services and supplies.
- Raw and packing materials lost in production.
- Energy.
- Depreciation, property taxes, and insurance.
- Cost of capital, that is, investment in land, buildings, machinery and equipment, inventories.

All of these elements add to the value of the raw materials being processed. Therefore, the value of the product equals the value of the raw materials plus the value of these input elements.

Since both the output data and the input data are converted to base-year dollars, the productivity index becomes a simple ratio of (1) the current productivity (output/input) stated in base-dollar terms and (2) the base-year productivity (output/input for the base). The base-year PI, of course, equals 1.0 since the base-year productivity figure would be divided by itself. If the productivity index for years subsequent to the base year is greater than 1.0, the productivity (output/input) for that year exceeded the base year productivity. Similarly, productivity deteriorated for the year if the PI for the year is less than 1.0.

The PI is a total-factor productivity index, that is, it includes all inputs that contribute to output. Many productivity indexes focus on one input, the most common being labor productivity, which commonly represents the production output (say, revenues or units produced) divided by the labor input (such as man-hours, number of employees, or even labor cost). Limiting the number of inputs, however, reduces the information conveyed by such partial productivity indexes. For example, a firm's labor productivity may be increasing while its capital productivity is decreasing, and unless the two are viewed together the net effect on the firm's profits cannot be computed.

As listed above, GF takes into account all input factors that add to the value of raw materials. Therefore, its PI reflects the net of productivity increases and decreases; that is, if labor productivity increases while capital productivity decreases by the same amount, the net effect on the PI is zero (assuming equal weighting of labor and capital). Since the inclusion of an input amount is dependent on whether the input is associated with the product output, the following explanation addresses first the output and then the input elements associated with the output elements.

III. Measuring Output—Base Year

To demonstrate the method of computing the PI, assume that a product requires two inputs—labor (including overhead and associated costs) and raw materials. To compute the output, GF determines for labor and materials the cost of a unit of output during the base year. For example, assume the following data:

COST ELEMENTS	BASE-YEAR COST
Labor cost/unit of output	$0.24
Materials-lost cost/unit of output	0.20

GF has this type of data available from its cost accounting system for each of its products. Basically, GF is interested in representing the output as the value added to the raw materials that are put into the production process. For example, how much more valuable is a pound of processed and canned coffee than a pound of raw coffee beans ready for processing? Notice that GF employs a materials-lost measurement for productivity measurement as opposed to a materials-used approach. There are two primary reasons for employing this approach:

• Materials lost represents the cost reduction opportunity at the plants. The purchasing of materials and the related price management is not a plant function but is carried out at the corporate level.

- Using materials lost reduces the impact of raw materials cost on the PI. Raw materials represent a high proportion of the cost of GF products. Changes in total materials cost caused by the volatility of commodity prices would distort the PI. Moreover, if materials-consumed data were included in the output figure, they would have to be eliminated in order to arrive at value added.

Since GF's cost accounting system generates a materials-lost figure, the data are readily available. An example of this lost cost approach is as follows.

Assume that during the base period 200,000 pounds of raw materials were lost while the plant was producing 500,000 cases of a finished product. If the base-year purchase price of the raw materials was $0.50, the total cost of lost raw material would equal $100,000 (200,000 × $0.50) and the unit cost would equal $0.20/case.

On the labor side of this example, assume that one employee making $12.00 an hour (including all overhead and related costs) produced 50 cases of product per hour (or $0.24/case) during the base year. Therefore, 500,000 cases would require a total labor cost of $120,000 ($0.24 × 500,000). If all of the raw material had found its way to the can, the value added to the raw material would equal the amount of labor necessary for processing (excluding, for this example, capital costs). However, when raw material is lost in production, the finished value must be increased to reflect the value of the loss. Therefore, the value added (output) during the base year is $220,000 ($100,000 in lost raw material cost and $120,000 in labor cost).

IV. Measuring Input—Base Year

In the above example, the inputs into the plant are raw materials and labor. As with the calculation of output, the materials input is represented by raw materials lost in production, which was given above as 200,000 pounds at $0.50/pound, or $100,000. The labor input portion is also identical to the output calculation ($120,000) since all labor input contributes directly to the value added to the raw material. The capital factor is not included in the calculation of the base-year PI since the capital productivity equals the difference in the current cost of capital and the cost of capital during the base year (making this difference zero when computing the base-period capital productivity). Therefore, the input total is also $220,000 ($100,000 for materials lost, $120,000 for labor, and zero for capital) and the PI for the base period is 1.0 ($220,000 for output divided by $220,000 of input). The PI of 1.0 is, of course, because the methods used to calculate (1) the value added to the raw materials (output) and (2) the total value of the input elements are essentially the same.

V. Measuring Output—Current Year

The PI for a year subsequent to the base year is computed by using current-year volume at base-year prices.* For example, if 600,000 cases of product were produced during the current year using 10,800 employee hours, the output (value added) would be $264,000: $120,000 in raw material (600,000 × $0.20/case) and $144,000 in labor (600,000 × $0.24/case). Again, notice that the prices used in computing the value added are base-year prices ($0.20/case for raw materials and $0.24/case for labor). The only variable, therefore, in the output calculation for the current year is the production volume.

VI. Measuring Input—Current Year

In computing the input (denominator) for the current-year PI, the current-year volume of input elements is used along with the base-period prices (purchase cost) of those inputs. In the above example, assume that during the current year 190,000 pounds of raw material were lost in production and that it took 10,800 labor hours to produce the 600,000 cases. The total input value (at base-year prices) for the current year would be $224,600: $95,000 for raw materials (190,000 units × $0.50/unit), $129,600 for labor (10,800 hours × $12.00/hour), assuming no change in the cost of capital.

The PI for the current year is, therefore, $264,000/$224,600, or 1.18. This calculation and the preceding steps are summarized in Table 6-1. The reason for using materials lost in production rather than materials used in production should be apparent from the data in Table 6-1. Notice how the materials portion of the total PI has a

---

*If base-year prices are not available, current-year prices are reduced by price deflators to estimate base-year prices. Although this technique is not preferable to using actual base-year prices, it is sometimes necessary, especially with pooled cost areas such as overhead and administration.

Table 6-1.   Calculation of Productivity Index.

OUTPUT

CATEGORY	BASE YEAR			CURRENT YEAR		
	TOTAL $ COST	UNITS PRODUCED	COST/UNIT	BASE-YEAR COST/UNIT	UNITS PRODUCED	TOTAL $ COST
Material (lost)	$100,000	500,000	$0.20	$0.20	600,000	$120,000
Labor	120,000	500,000	0.24	0.24	600,000	144,000
Total	$220,000					$264,000

INPUT

CATEGORY	BASE YEAR			CURRENT YEAR		
	UNITS OF INPUT	PURCHASE COST/UNIT	$ COST OF INPUT	BASE-YEAR PUR. COST/UNIT	UNITS OF INPUT	$ COST AT BASE-YEAR PRICES
Material (lost)	200,000	$ 0.50	$100,000	$ 0.50	190,000	$ 95,000
Labor	10,000	12.00	120,000	12.00	10,800	129,600
Total			$220,000			$224,600

BASE-YEAR PI

$$\frac{\text{Output}}{\text{Input}} = \frac{\$220,000}{\$220,000} = 1.0$$

CURRENT-YEAR PI

$$\text{Material PI} = \frac{\$120,000}{\$\ 95,000} = 1.26$$

$$\text{Labor PI} = \frac{\$144,000}{\$129,600} = 1.11$$

$$\frac{\text{Total}}{\text{Factor PI}} = \frac{\$264,000}{\$244,600} = 1.18$$

somewhat lesser weight than the labor portion. Had the materials calculation considered materials used instead of materials lost, the materials would have had a much greater weight than labor in the total PI calculation, with the danger that a disproportionate cost reduction effort might be directed by the plant to materials rather than to labor and materials equally.

One of the principal advantages of this method of calculating the PI is its ability to absorb without distortion the effect of changes in product mix. For example, as shown in Table 6–2, the plant produces three products in unequal proportion from year to year. In spite of the change in product mix, however, the 1974 PI is still 1.0 (with 1973 as the base year), and in 1975 the PI of 0.83 reflects a reduction in

Table 6-2.   Raw Materials Lost in Production. Product Mix Example, Fiscal 1973, 1974, 1975.

OUTPUT

PRODUCT	STD. UNITS PRODUCED			F1973	OUTPUT: VALUE ADDED		
	F1973	F1974	F1975	COST/STD. UNIT	F1973	F1974	F1975
A	1,000	3,000	10,000	$ 1.00	$ 1,000	$ 3,000	$10,000
B	2,000	2,000	0	5.00	10,000	10,000	0
C	3,000	1,000	0	10.00	30,000	10,000	0
Total					$41000	$23000	$10000

INPUT

PRODUCT	QTY. LOST IN PRODUCTION			BASE-YEAR PURCHASE	INPUT AT BASE-YEAR PRICES		
	F1973	F1974	F1975	STD. UNIT	F1973	F1974	F1975
A	2,000	6,000	24,000	$ 0.50	$ 1,000	$ 3,000	$12,000
B	4,000	4,000	0	2.50	10,000	10,000	0
C	6,000	2,000	0	5.00	30,000	10,000	0
Total					$41000	$23000	$12000

Partial Productivity Index—Raw Materials Lost in Production	1.00	1.00	0.83

Table 6-3.   Plant A Output Versus Input ($ millions).

	1973	1974	1975	1976	1977	1978 PLAN
OUTPUT						
Raw materials lost in production	$42.6	$41.9	$33.2	$39.8	$35.5	$35.3
Packing materials lost in production	0.3	0.3	0.3	0.3	0.3	0.3
Labor and overhead	25.4	25.5	19.2	22.7	19.9	19.4
Total output	$68.3	$67.7	$52.7	$62.8	$55.7	$55.0
INPUT						
Raw materials lost in production	$42.6	$41.0	$32.4	$37.9	$34.2	$32.7
Packing materials lost in production	0.3	0.3	0.3	0.4	0.3	0.3
Labor and overhead	25.4	25.3	26.2	28.1	25.3	25.9
Increase in cost of capital	—	—	0.4	(0.9)	(1.0)	(1.4)
Total input	$68.3	$66.6	$59.3	$65.5	$58.8	$57.5
Productivity index (output/input)	1.000	1.017	0.888	0.958	0.947	0.956
Volume (millions of units)	18.1	17.6	14.5	16.2	13.5	12.5

materials productivity (a greater percentage of materials are lost) during 1975 and not the effect of a shift in product mix.

VII. Translating Productivity Indexes into Cost Reduction Targets

Table 6-3 displays the basic data for computing the PI for six years including the base year 1973 and a projected PI for 1978. In addition to the two basic elements (labor and materials) the data include a line for packaging materials. Notice further that the cost-of-capital factor is contained only among the inputs and there only as a change in the cost of capital from the base year. The production volume figures are included to permit an analysis of the PI as a function of the throughput of the plant.

The data from Table 6-3 provide some initial insight into why the productivity of the plant has fluctuated. For example, it appears that volume reductions have had a significant negative impact on the absorption of fixed costs. In this respect the plant should be more responsive to volume declines in incurring fixed costs. It also appears that the unfavorable trend in variable cost productivity (primarily labor) is partially a result of low volumes. The reduced raw material shrinkage and reduced cost of capital have been the brightest spots over the last six years. Finally, as explained below, it will take a reduction in input costs of $7.5 million (or 13 percent of the $57.5 million) to attain a target index of 1.10 in 1978.

As suggested by the mention of a target index, another benefit of the General Foods Productivity Index calculation is that it permits the computation of a bottom-line cost reduction figure that can be associated with a particular productivity index and that can be given a plant manager as a cost reduction target. For example, if Plant A (Table 6–3) were to establish a PI objective of 1.0 for the year 1978, it would need to reduce its input cost by $2.5 million, since the value added (output) in 1978 is expected to be $55 million and the input is expected to be $57.5 million. Table 6–4 depicts the plant manager's plan to realize the cost reductions necessary to reach a target PI or 1.0. A similar analysis can be used to verify the statement above that a $7.5 million reduction in input is needed to attain a target index of 1.10.

Another example of the productivity trend for a plant is given in Table 6–5. In this case, Plant B appears to have experienced a positive change in productivity over the years with a few dips due to volume reduction. The primary improvements are in the area of raw material yield and to a lesser extent in labor productivity. There are no significant unfavorable trends. Table 6–6 outlines how the manager of Plant B anticipates raising his PI to 1.123 by saving $720,000, that is, by reducing his inputs by that amount.

VIII. Application of PI at Division Level

The PI calculation is not restricted to use at the plant level. It can also be usefully employed at the division level where the plant PIs are

Table 6-4. Plant A Plant Productivity Index.
Impact of 1978 Cost Reduction Program ($ thousands).

PROJECT TITLE AND DESCRIPTION	ANNUAL NET SAVINGS (1978 $)	ANNUAL NET SAVINGS (1973 $)	COST OF CAPITAL (F1973 $)	$ CHANGE IN PLANT INPUT
Equipment replacement	$3492	$1088	$ 78	$1010
Reduction of raw material losses	1700	527	69	458
Energy Conservation	250	24	—	24
Crewing	350	241	—	241
Reformulation	1043	717	—	717
Other opportunities	161	111	—	111
Total plant	$6996	$2708	$ 147	$2561

1978 Productivity Index: 0.956
Revised 1978 index if all cost reduction is achieved: 1.001

Table 6-5.   Plant B Output Versus Input ($ millions).

	1973	1974	1975	1976	1977 L/E	1978 PLAN
OUTPUT						
Raw materials lost in production	$22.7	$22.6	$20.5	$23.8	$24.6	$23.2
Packing materials lost in production	0.2	0.2	0.2	0.2	0.2	0.2
Labor and overhead	11.4	12.0	10.9	12.6	12.7	12.1
Total output	$34.3	$34.8	$31.6	$36.6	$37.5	$35.5
INPUT						
Raw materials lost in production	$22.7	$22.2	$19.9	$22.3	$22.2	$21.3
Packing materials lost in production	0.2	0.2	0.2	0.2	0.2	0.2
Labor and overhead	11.4	11.3	10.9	11.9	12.1	11.9
Increase in cost of capital	—	(0.4)	(0.3)	(0.8)	(0.9)	(1.1)
Total input	$34.3	$33.3	$30.7	$33.6	$33.6	$32.3
Productivity index (output/input)	1.000	1.045	1.029	1.090	1.112	1.098
Volume (millions of units)	18.5	18.7	17.1	19.9	19.9	19.0

Table 6-6.   Plant B Plant Productivity Index.
Impact of 1978 Cost Reduction Program ($ thousands).

PROJECT TITLE AND DESCRIPTION	ANNUAL NET SAVINGS (1978 $)	ANNUAL NET SAVINGS (1973 $)	COST OF CAPITAL (F1973 $)	$ CHANGE IN PLANT INPUT
Packing material specification changes	$ 80	$ 49	$ (8)	$ 41
New blending system	45	24	(11)	13
Vacuum pump	20	7	(1)	6
Recirculation of cooling water to vacuum pumps	13	4	(2)	2
Agglomerator	20	7	(2)	5
2-lb. Lid applicator	30	19	(2)	17
New scale—can line no. 1	120	37	(9)	28
Yield improvement of 1.0% on product X	670	214	—	214
Yield improvement of 0.8% on product Y	232	74	—	74
Yield improvement of 0.2% on product Z	134	43	—	43
Sewer connection charge not needed	200	138	—	138
Reduce efficiency loss on processing equipment to 4%	150	103	—	103
Other opportunities	86	36	—	36
Total plant	$1800	$755	$ (35)	$720

1978 Productivity Index: 1.098
1978 index if all cost reduction is achieved: 1.123

aggregated into a division PI. These data are reflected in Table 6–7 in the same format as the individual plant data, and as with the plant data, the division data can provide valuable insights on division operations. For example, the primary reasons for the maintenance of the division PI at the base-year level is that improved raw materials yields and reduced capital investment have offset the drop in labor productivity and the underabsorption of fixed overhead over the years. Reduced volume and its resultant unfavorable effect on fixed cost absorption has been the single most negative factor.

Just as the plants can have PI targets translated into cost reduction targets, so can divisions establish cost reduction objectives as shown in Table 6–8. Table 6–9 depicts a summary of the cost reductions necessary from each plant in a division in order for the division to meet the PI objectives for 1978 and 1983.

IX. Role of Plantwide Productivity Measurement Program at GF

The primary purpose of the PPMP is to assist in meeting the company's continuing goal of General Foods being the low-cost producer in the food industry. Some of the uses of PPMP are:

- To permit analyses of operations by cost element (raw materials, labor, and capital).
- To motivate the plant managers in achieving cost reduction.
- To provide a common measure between plants and between divisions.
- To permit the measuring of the effects of cost reduction programs.
- To provide a basis for long-range and short-range planning to include facilities planning.
- To identify areas of potential improvement.
- To permit the evaluation of various engineering techniques for cost reduction.
- To highlight areas (by subelement) of potential improvement.

As with any good system, the PPMP is being constantly studied and refined. As more plants participate in the system, it is likely to gain in acceptance companywide. The motivation provided by the individual plant productivity indexes is sufficient by itself to justify the system.

Table 6-7.   Division Productivity Measurement ($ millions).

	1973	1974	1975	1976	1977 L/E	1978 PLAN
OUTPUT						
Raw materials lost in production	$121.3	$122.1	$107.3	$116.5	$102.4	$ 96.5
Packing materials lost in production	1.0	1.1	1.0	1.0	0.9	0.8
Labor and overhead—plants	79.3	83.6	72.4	78.6	68.5	64.5
Labor and overhead—div. HQ operations	2.2	2.3	2.0	2.2	1.9	1.8
Total output	$203.8	$209.1	$182.7	$198.3	$173.7	$163.6
INPUT						
Raw materials lost in production	$121.3	$119.6	$104.6	$112.6	$ 97.2	$ 89.5
Packing materials lost in production	1.0	1.0	0.9	1.1	0.9	0.8
Labor and overhead—plants	79.3	82.5	83.0	87.3	78.9	77.8
Labor and Overhead—div. HQ operations	2.2	1.6	1.5	1.6	1.9	1.9
Increase (decrease) in cost of capital over base year—plants	—	(0.7)	0.4	(3.1)	(3.4)	(4.7)
Increase (decrease) in cost of capital over base year—div. HQ operations	—	(0.5)	(1.0)	(2.0)	(1.2)	(2.0)
Total input	$203.8	$203.5	$189.4	$197.5	$174.3	$163.3
Productivity index (output/input)	1.000	1.027	0.964	1.004	0.997	1.002

Table 6-8. Division Summary, Plant Productivity Index. Impact of 1978 Cost Reduction Program ($ thousands).

PLANT	ANNUAL NET SAVINGS (1978 $)	ANNUAL NET SAVINGS (1973 $)	LESS COST OF CAPITAL (1978 $)	$ CHANGE IN PLANT INPUT
Plant A	$ 6,996	$ 2,708	$147	$ 2,561
Plant B	1,800	755	35	720
Plant C	4,301	2,373	28	2,345
Plant D	414	219	29	190
Total division	$13,511	$ 6,055	$239	$ 5,816

PLANT	1978 INDEX BEFORE C/R	INDEX CHANGE DUE TO C/R	REVISED INDEX INCLUDNG C/R
Plant A	0.956	0.045	1.001
Plant B	1.098	0.025	1.123
Plant C	0.953	0.041	0.994
Plant D	1.093	0.013	1.106
Total all plants incl. div. HQ	1.002	0.037	1.039

Table 6-9.   Attainment of Division Objectives ($ thousands).
Assuming that a division's objective is to achieve a productivity index of 1.10 in 1978 and 1.20 in 1983, each plant must make some significant progress in reducing its level of input. The following example shows for each plant and the division the reduction in input needed to attain these objectives.

| | DIVISION OBJECTIVES | |
	1978 COST REDUCTION	1983 COST REDUCTION
Plant A	$ 7,526	$11,692
Plant B	58	2,749
Plant C	7,573	11,661
Plant D	541	1,412
Total division	$15,698	$27,514

However, the additional uses of the system in planning and analysis have made the PPMP a very valuable management tool.

General Foods has realized that as world food problems grow, members of the food industry will be under increasing pressure to process and distribute food products using the most efficient methods available. Its management believes that in the PPMP it has a meaningful measure of its progress toward optimum efficiency in the manufacturing process. Its general-purpose design permits application of the system on the myriad of food products already processed by GF as well as on future products. Its design simplicity also allows for translating productivity indexes into specific cost reductions, that is, into a language that plant managers understand and are responsive to. The PPMP, therefore, combines the attributes of a useful and flexible tool for corporate management with an understandable and accepted measure for the man at the plant.[8]

REFERENCES

1. "Distribution Illustrated," *Handling and Shipping,* May 1980.
2. Dallos, R.E., "Airlines Put Planes on a Diet," *San Jose Mercury,* June 23, 1981.
3. "Amway TILTS Productivity," *AMAGRAM,* Amway Corp., Winter 1981.
4. *San Francisco Sunday Examiner and Chronicle,* May 10, 1981.
5. Western Electric Co., *Newsbriefs Background,* June 9, 1981.
6. Central Methodist College, *Central Bulletin,* April 1981.
7. Western Electric Co., "Cost Reduction Manual," Hawthorne Works, 1979.
8. American Productivity Center, "Measuring Productivity at the Firm Level," December 1980.

7
Program Development

A program that fits your company is essential in the quest for cost reduction. In general, all programs will have one common link that is necessary for the program to succeed. The common link can be defined as open communication.

Successful cost reduction programs can usually be associated with effective communication between the various groups that supply input ideas. The idea input may come from management, engineering, or other segments of the labor force. Once ideas have been developed the next task is that of reviewing each approved idea at a scheduled meeting. In broad terms the weekly or monthly cost reduction meeting can be described as the forum for processing cases in an orderly manner. Attendance by committee members and guests is important and should be given priority over routine matters that may arise on a periodic basis.

STARTING YOUR OWN PROGRAM

How can you start your own program? Your best approach is to discuss the concept with your supervisor and enlist his support. He should be able to provide feedback on the current company climate regarding this endeavor.

At the same time your supervisor is evaluating the new venture into a systematic program, you can be working in the following areas:

1. Develop a proposed charter that would cover an expanded role into cost reduction/productivity involvement.

2. Define productivity progress in your company. You will probably need some assistance from accounting in this area. Review the

last annual report, compare your company with others that compete in the same marketplace. Productivity is doing more with the resources we have. It is measured as the simple ratio:

$$\frac{\text{Output}}{\text{Input}}$$

Output   refers to goods and services produced and may be expressed as dollars, pounds, etc.

Input   refers to labor, material, capital, etc.

3. Define productivity improvement. Productivity improvement is not just working longer and harder, it is working smarter. It means devising a method to get the best return on our investment in people, raw materials, facilities, equipment, and other resources. Some of the ways to improve productivity include:

• Creating a more productive working environment; reviewing people's concerns and seeking inputs for improvement.
• Developing new products and technology that challenge the marketplace and result in lower costs.
• Managing our production resources more effectively; developing action review audits.
• Utilizing machinery and equipment more efficiently; auditing utilization.
• Improving our working methods; conducting random audits.
• Motivating and training our people; showing a genuine concern.

In short, we produce more when we have better tools, production techniques, and are better organized and managed.

4. Develop awareness of cost reduction. To be effective, every segment of the company should understand its responsibility for improving productivity. This understanding is communicated through various methods. Among these are the following:

• Establish productivity goals and cost reduction objectives.
• Develop productivity measures for the shop, warehouse, and office.
• Encourage participation of all employees in the cost reduction improvement effort.
• Evaluate cost reduction training programs.

Let us break each of these steps down a little further.

5. Establish goals and objectives. To be successful a cost reduction improvement effort must include productivity goals. Try some goals even if you are not sure they are right; make corrections later. The key is to establish reaching goals.

The following is a simple five-step technique for cost reduction goal setting:

- Decide how to monitor achievement.
- Determine your productivity level.
- Set attainable goals with cost reduction conversions.
- Establish action plans for improvement.
- Implement cost reduction improvements.

Once you have decided on the output-to-input ratios that best fit your work, set improvement targets for the next month or year.

6. Develop productivity measures. To be meaningful the productivity measure used should relate to the organization unit. The following are measures that might be considered:

- *Employee productivity,* measured as output per employee or per man-hour, such as:
  —dollar sales per employee
  —value added per payroll dollar
  —units per man-hour
- *Equipment productivity,* measured as output per machine hour, such as:
  —pounds per machine hour
  —units per machine hour
  —machine utilization or downtime
- *Assets productivity,* measured as output per asset dollar, such as:
  —sales per asset dollar
  —units per asset dollar
- *Energy productivity,* measured as product output per energy input, such as:
  —units of product per cubic feet (gas)
  —units of product per gallons (oil)
  —units of product per kilowatt-hour

7. Encourage participation of all employees. The emphasis on productivity/cost reduction as a company goal comes from the top. But the best ideas for improving operations often come from the bottom. The involvement of people at all levels will ensure successful efforts.

The following techniques, communicated properly, can provide a participative atmosphere for productivity improvement:

- Encourage employee suggestions.
- Encourage employee participation in cost reduction.
- Consider employee recognition methods such as:
  —employee productivity awards
  —cost reduction contests
  —gift certificates for achievement
- Use employee motivation techniques such as:
  —employee suggestion awards
- Monitor employee productivity.

8. Evaluate cost reduction training programs. Many techniques are available to engineers, managers, and supervisors that can aid in achieving cost reduction goals:

- *Work simplification* can be used in office, shop, and warehouse.
- *Short-interval scheduling* can be used to control investment.
- *Employee motivation methods* can be utilized to generate input ideas.
- *Improvement of supervisory effectiveness* can be useful in goal setting.
- *Value analysis* can be applied to products and services.
- *Automation applications* can be applied if justified by rate of return.
- *Methods analysis* can be applied in all operational functions.

AFTER THE PROGRAM IS DEFINED

Once the program is defined each manager, supervisor, and engineer has a responsibility to improve productivity within his or her functional area. The methods used to improve productivity should be tailored for each situation. Successful programs usually involve positive employee attitudes. It is a management function to perceive present attitudes and work to improve them. Engineers have the challenge of improving technical systems and the related quality measurement techniques. My experience has been that the written case method of cost reduction is one of the most useful approaches that can be utilized. The written case method can be developed from a logical sequence of events. In broad terms, the first item in the chain of events is the development of an idea that appears to have potential. Potential can be defined as a projected dollar savings that is related to a product or service. Each item that is deemed to have savings potential should be documented for investigation. Refer to Figure 7-1 for an example of the steps involved in processing an idea.

Premeeting Idea Development

Before any cost reduction case can contribute to the bottom line of any product line a lot of groundwork has taken place. Idea conception is the most vital link in a chain of events. Every idea can be handled in an orderly manner if a stepping-stone approach such as a well-defined cost reduction program is utilized. Once the idea is documented the next step is to determine the feasibility of application. This part of the case development can be done by the individual who conceived the idea or perhaps someone who has a special knowledge in a particular field. The approach may follow the steps listed below:

- Investigate and assemble data.
- List differences in present versus proposed method.
- Define advantages of the new method.
- Determine requirements for labor and the associated labor rates.
- Determine current production volumes and the trend for the next five years.
- Compute an estimate of first-year savings and a five-year summary by year.

Assuming that the feasibility study supports your idea, where do we go from here? Depending on the paperwork flow in your cost reduction program, the next step may be your presentation at the meeting. My experience is that a scheduled meeting on a monthly basis is adequate. In the first phase of starting a new program weekly meetings may be beneficial, and even a requirement.

```
┌─────────────────────────────────┐      ┌─────────────────────────────────┐
│           BASIC IDEA            │      │   INVESTIGATION AUTHORIZATION   │
│                                 │      │   Brief Title                   │
│  1.  Present Method             │      │   Scope and Subject             │
│                                 │      │                                 │
│  2.  Proposed Method            │      │                                 │
│                                 │      │               Complete Date     │
│  3.  Estimated Savings          │      │               Rate of Return    │
│                                 │      │               Make vs. Buy      │
│  4.  Signature                  │      │                                 │
│                                 │      │               Control Number    │
│  Attachments ____Yes ____No     │      │                                 │
│                                 │      └─────────────────────────────────┘
└─────────────────────────────────┘

┌─────────────────────────────────┐      ┌─────────────────────────────────┐
│  OUTLINE OF WORK AND SCHEDULE   │      │   ESTIMATED COST AND SAVINGS    │
│  1. _____ _____     │      │  1.  Development Expense - $     │
│  2. _____ _____     │      │  2.  Associated Expense  - $     │
│  3. _____ _____     │      │  3.  New Plant Items     - $     │
│                                 │      │  4.  Total (1, 2, 3)     - $     │
│  Suggested By   Conducted By    │      │  5.  Annual Savings             │
│  Reason For Reissue             │      │        Current Year      - $     │
│                                 │      │        Est. 5-Year Avg.  - $     │
│                                 │      │  6.  Disposal of Plant          │
│                                 │      │        Cost              - $     │
│                                 │      │        Cost of Disposal  - $     │
│                                 │      │        Salvage Value     - $     │
│                                 │      │  7.  Approvals:                 │
└─────────────────────────────────┘      └─────────────────────────────────┘
```

Figure 7-1. Processing an idea.

Published Schedule of Meetings

A published schedule is a must in order to assure that everyone is aware of two critical dates: the cutoff date and the meeting date.

• The cutoff date is defined as the last day that a case can be submitted to the accounting group for review. Only cases submitted by this date can be placed on the agenda for the monthly meeting. What types of cases are included in this review? The answer: all types of cases that come before the committee. This includes all new cases, cases that are reissued for a specific reason, and the real meat of the program, all cases that are being closed with specified dollar savings. This point will be expanded further with examples in the chapter.

• The meeting date is the scheduled date of the cost reduction meeting. As a general rule the meeting dates are scheduled for the entire year. This is essential for the participants and guests to plan their arrangements. This becomes even more important when people from several plant locations are part of the committee. Regardless of how your program is designed to work, scheduled meetings are a must. Refer to Figure 7-2 for an example of a typical schedule that is published in advance of all meetings.

Committee Members

Any organized involvement—baseball, football, basketball, cost reduction—is composed of people who play certain roles. In the case of the monthly cost reduction meeting, the chairman is the captain of the team. The chairman of the committee is usually from the engineering group. In most cases he or she is usually the ranking member of management on the cost reduction committee.

1. *The chairman,* in broad terms, is responsible for the following:

• To define the goals and objectives of the program.
• To provide ongoing direction for the cost reduction committee.
• To select the coordinators for the various product lines of business or in some cases representatives for remote plant sites.
• To review monthly achievement related to predetermined goals.
• To keep upper management informed of progress as well as problem areas when achievement of goals appears to be lagging.
• To publicize the achievement of the cost reduction committee.

December 22, 1980

Memorandum for Record

The following information is furnished for your planning in conjunction with the 1981 program. Cases to be included in the Cost Reduction meeting agenda must be in Accounting ten days prior to the meeting per the following schedule:

Cutoff Date	Meeting Date
	01/13/81
01/20/81	02/03/81
02/17/81	03/03/81
03/24/81	04/07/81
04/21/81	05/05/81
05/18/81	06/02/81
06/22/81	07/07/81
07/21/81	08/04/81
08/08/81	09/01/81
09/22/81	10/06/81
10/20/81	11/03/81
11/13/81	12/01/81

Please distribute this information to all parties concerned. If you have any questions, please contact me.

Secretary

Copy to:
All Cost Reduction Coordinators

Figure 7-2. Memorandum for record.

2. *Cost reduction coordinators* are the real work horses of the committee. The success of any cost reduction program depends on selling ideas, and this can only be achieved through effective communication. When people are selected to fill this function the following should be considered:

- The coordinator must be skilled in communications; this includes both verbal and written forms.
- The coordinator must be able to deal with other engineers and stimulate the flow of new ideas into the program.
- The coordinator must be able to deal with all levels of management and utilize visual media to document accomplishments.
- The coordinator must be respected by his peers. This respect is earned from previous successes in the program. In some organi-

zations a success may be defined as the achievement of $1.0 million in cost reduction savings.

3. *The cost reduction secretary* may be a member from the accounting group. This seems like a logical selection since all paper flows through this group at least ten days before the meeting date.

This interval allows time to review each case for neatness, originality, and aptness of thought. Also, production volumes and projected savings can be reviewed for any extension errors.

The secretary also records any vital transaction that takes place during the meeting. At the meeting the secretary passes out a copy of the agenda and each case that is to be reviewed.

4. *Cost reduction conductors* are the engineers who attend specific meetings and are not members of the official committee. Their function is to report on a specific action plan that is to be followed or to report on a completed investigation case that has been wrapped up with a specific amount of dollar savings. The use of samples, photographs, slides, or other visual means can enhance any cost reduction presentation. This is very important for other members of the committee who may not be as familiar with the case as the conducting engineer.

5. *Selected guests* are usually members from upper management. Their attendance adds to the meeting in that it displays the company's ongoing interest in the program. Their presence also allows them to pass along other items of interest that may not have been filtered down through the normal channels. The meeting allows management to inject feedback on earnings, sales, and a projection of the business picture for the future. From time to time, if things are lagging in the results achievement—the savings that were planned for a given point in time—a few well-placed words always seem to motivate the committee and others who contribute input information.

Location of Meetings

If your program is conducted in a company that has only one plant location or distribution site, the meeting would normally be held at that location. However, if the cost reduction program you are now associated with or wish to start is to be spread over several states and multiplant sites, you have a different problem: maintaining an effective communication network. In my experiences, I have seen the pattern for selecting meeting locations go full circle.

(Ⓐ) Western Electric

Investigation Case Authorization

(411,078-203)

CASE NO. 411,083

ISSUE NO. 2

PAGE 1 OF 2 PAGES

FUNCTION : WHSE

PECC/RECC: WHSE

KEY WORD INDEX

1. CONSOLIDATE

2. PLUG-IN

3. POOL

TITLE (BRIEF DESCRIPTION)

CONSOLIDATION OF PLUG-IN POOL

SCOPE AND OBJECT

THIS CASE PROPOSES TO RELOCATE THE NCN AREA PLUG-IN POOLS NOW STORED IN THE SAN LEANDRO POLVOROSA STREET AND SACRAMENTO SIXTH STREET LOCATIONS, TO THE SAN LEANDRO VERNA COURT DRIVE LOCATION. (1)

THIS CONSOLIDATION WILL RESULT IN LABOR AND OPERATING COSTS SAVINGS.

LOCATION   NCNSC

ORGANIZATION NO. 22PC818250

PROBABLE COMPLETION DATE   8/82

CLASSIFICATION (SEE APPLICABLE INSTRUCTION)

[X] COST REDUCTION   [ ] DEFINITE ROUTINE

[ ] OTHER DEVELOPMENT [ ] GENERAL ROUTINE

[ ] BELL SYSTEM SAVINGS CASE

ASSOC. PLANT AUTH. NO. _____

DATE APPVD. BY C.R. COMM. _____

RATE OF RETURN _____

MAKE VS. BUY APPROVAL _____

OUTLINE OF WORK AND SCHEDULE

OPEN INVESTIGATION CASE	2/81
CONDUCT FEASIBILITY STUDY	6/81
CONSOLIDATE LOCATIONS	5/82
CLOSE CASE (2)	8/82

LOCATION DIVISION OF SAVINGS

100% TO NCNSC

NO APPLICATION AT SCSC & WIOSC

ESTIMATED COST AND SAVINGS DATA

1. DEVELOPMENT EXPENSE	
(A) ENGINEER'S SALARY LOADED	$2,496.00
(B) EXPERIMENTAL SHOP WORK	0
(C) EXPERIMENTAL PLANT OR DEVELOPMENT FACILITIES	0
TOTAL	$2,496.00
2. ASSOCIATED EXP. TO BE INCURRED AS A RESULT OF THIS CASE	
(A) MOVES AND REARRANGEMENTS (3)	$12,000.00
(B) REMODELING PLANT	0
(C) EXPENSE TOOLS AND SUPPLIES	0
(D) ENGRG. SERVICES	0
(E) SERVICES, OTHER THAN ENGRG. (4)	$93,000.00
TOTAL	$105,000.00
3. NEW PLANT REQUIRED AS A RESULT OF THIS CASE (EXCLUDING ITEM 1 (C)) (A) LAND, BLDGS., LAND IMPRV., MACH., & TRANSP. EQUIP. (INCL. DESIGN)	0
(B) SMALL TOOLS	0
(C) FURN. AND FIXT. (5)	$1,000.00
TOTAL	$1,000.00
4. TOTAL EXPENDITURES (ITEMS 1,2, & 3)	$108,496
5. ANNUAL SAVINGS BASED ON	
(A) 5 YEAR AVERAGE	$225,796
(B) CURRENT LEVEL*	$192,442
COST ESTIMATE NO. _____ DATE _____	

APPROVED BY: _____ ORG. NO. _____

* 12 MONTHS SUBSEQUENT TO THE DATE SAVINGS ARE MADE EFFECTIVE

	SUGGESTED BY	CONDUCTED BY
W. E. RINGEN - SV		
G. C. GEBHARD - SV	E. A. CRINER - SV	
E. A. CRINER - SV		

REASON FOR REISSUE (1) LOCATION CHANGE.

(2) TO CHANGE CLOSING DATE.

(3) TO INCREASE COST OF MOVING MATERIAL.

(4) COST TO LEASE RACKS AND BUILDING LEASE DIFFERENCE.

(5) MISC. TAPE MACHINES AND HAND STRAP DEVICES.

6. PLANT TO BE REPLACED

	LAND, BLDGS. LAND IMPRV. MACH. & TRANSP. EQUIP.	SMALL TOOLS FURN. & FIXT.
(A) COST		
(B) COST OF DISPOSAL		
(C) SALVAGE VALUE		

APPROVALS:

PRELIM. _____

DIST. ENGR. _____   C/R COORD. _____   DEPT. CHIEF ENGRG. _____

DEPT. CHIEF ENGR. COORD _____

FINAL _____   DATE _____

Figure 7-3.   Investigation case authorization.

In a small organization the location of the meeting will be no problem. It will usually be held in the same facility. A larger organization with more than one operation such as a sales office, manufacturing plant, or warehouse may choose to vary the meeting location in order to provide an opportunity for exposure to the meeting and how the system works.

Your Monthly Meeting

The monthly meeting will deal with new ideas that are coming before the committee for the first time. Also, cases that have been previously approved must be reexamined when the original scope is changed. A typical example of such a case is shown in Figure 7-3.

SUPPORTING DETAILS

DEVELOPMENT EXPENSE
ENGINEER's SALARY $24.96 × 100 HR = $2,496.00

ASSOCIATED EXPENSE
ESTIMATED RELOCATION EXPENSE = $ 12,000—MOVEMENT OF MATERIAL
$ 30,000—FOR RACKING LEASE
$ 63,000—FOR ADDITIONAL BUILDING
LEASE COSTS
$105,000

NEW PLANT
FURNITURE AND FIXTURES = $1,000—FOR TAPE MACHINES.

SAVINGS
IT IS ESTIMATED THE CONSOLIDATION OF THE PLUG–IN POOL AT ONE LOCATION WILL ELIMINATE SIX GRADE 3 WAREHOUSE EMPLOYEES.

CURRENT LEVEL
$15.42/HR × 40 HR/WK × 52 WK/YR × SIX EMPLOYEES = $192,442.00

FIVE-YEAR AVERAGE (ASSUME 8% LABOR INCREASE)
1ST YEAR $ 192,442
2ND YEAR $ 207,837
3RD YEAR $ 224,464
4TH YEAR $ 242,421
5TH YEAR $ 261,815

$1,128,979 ÷ 5 = $225,796

Figure 7-3. (continued)

Each change is listed in the block "reason for reissue." In order to provide the complete picture the entire case must be rewritten. This includes the investigation case authorization, supporting details, and the return on investment analysis. Return on investment is where the

COST REDUCTION ESTIMATE CONSOLIDATION OF PLUG-IN POOL PRINT 06/09/81
FILE: CRINER /1981   UPDATE 06/09/81

	NET COST OF CASE	INCOME TAX CREDIT RATE	NET AMOUNT
LINE	(A)	(B)	(C)
1. BUILDING AND LAND IMPROVEMENTS	0.0	XX	XX
2. OTHER PLANT	1000.0	8.60%	86.0
3. INVESTMENT TAX CREDIT	XX	6.67%	66.7
4. ASSOCIATED EXPENSE	105000.0	43.00%	45150.0
5. DEVELOPMENT EXPENSE	2496.0	43.00%	1073.3
6. PRODUCT INVENTORY	0.0	XX	XX
7. NET SALVAGE	0.0	XX	XX
8. TOTAL	108496.0	XX	XX

	YEAR	NET SAVINGS GROSS * 0.5700	INCOME TAX CREDIT PLANT	EXPENSE	NET CASH RETURN
	(D)	(E)	(F)	(G)	(H) = E + F + G
9.	1	109691.9	86.0	46290.0	156067.9
10.	2	118467.1	86.0	XX	118553.1
11.	3	127944.5	86.0	XX	128030.5
12.	4	138180.0	86.0	XX	138266.0
13.	5	149234.6	86.0	XX	149320.6
14. TOTAL		XX	XX	XX	690238.1

	YEAR	PRESENT VALUE AT	129.4% RATE OF RETURN
	(I)	(J)	
15.	1	68027.7	
16.	2	22524.6	
17.	3	10603.0	
18.	4	4991.2	
19.	5	2349.5	
20. TOTAL		108496.0	

21. AVERAGE ANNUAL RETURN (14H/5 YEARS) = $138047.6

22. COST RETURN RATIO (8A/LINE 21) =   0.7859

23. RATE OF RETURN = 129.4%

24. PREPARED BY: E A CRINER

Figure 7-4.   Rate of return worksheet.

total picture comes into focus. Refer to Figure 7–4 for a typical analysis.

Once a study has been completed an investigation report is issued. See Figure 7–5. This report can be designed to fit any situation. The illustration has worked well and can be modified to suit your needs. The

REPORT ON INVESTIGATION CASE

SUBJECT. _____   CASE NO. _____

                                              LOCATION _____

CONDUCTED BY _____   DATE   _____

SUGGESTED BY _____

SCOPE AND OBJECT: A BRIEF STATEMENT OF THE PURPOSE AND OBJECTIVES.

ACTION TAKEN: A BRIEF SUMMARY OF WORK PERFORMED.

RESULTS OBTAINED: A SUMMARY OF RESULTS OBTAINED.

PATENT CONSIDERATIONS: DOES THE CASE HAVE ANY?

EXPENDITURES AND SAVINGS: A SUMMARY OF BOTH.

EXPENDITURES

DEVELOPMENT EXPENSE	ASSOCIATED EXPENSE	PLANT
ESTIMATED ACTUAL	ESTIMATED ACTUAL	ESTIMATED ACTUAL
$_____ $_____	$_____ $_____	$_____ $_____

SAVINGS

	ESTIMATED	ACTUAL
CURRENT YEAR	$_____	$_____
FIVE YEAR AVERAGE	$_____	$_____

RATE OF RETURN: _____ %

APPROVALS

INVESTIGATING ENGINEER _____   CHAIRMAN _____

DEPARTMENT CHIEF, PRODUCTION _____   DEPARTMENT CHIEF, ACCOUNTING____

DEPARTMENT CHIEF, ENGINEERING_____   PLANT MANAGER _____

Figure 7–5.   Report on investigation case.

results can be verified by Accounting to insure that savings claimed are real and will stand up under close scrutiny.

Processing of Cost Reduction Cases

The flowchart in Figure 7–6 illustrates in general the action steps that are required in the processing of an idea associated with cost reduction.

Once the idea has been documented as a proposal it is routed to the local coordinator for review. If the concept is not currently under investigation the official case can be developed. At the monthly meeting the case is presented to the committee. Once the case is accepted work can be started. At this point in time the following events also start:

- Time charges are recorded and charged to the assigned case number. This is shown on the weekly time sheet.
- Associated expenses are also recorded against the authorized case number.
- Plant or capital expenditures are also recorded against the case.
- As a general rule these factors could be recorded by the local accounting group on a weekly basis or as they occur.

These factors should be watched carefully by the engineer who is working on the case. When it appears that the approved limits will be exceeded for some reason, the case should be reissued for management's approval.

What does this mean? Simply stated it means that the scope of the case has changed; this could be due to higher dollar expenditures or a delayed closing date. Cases should be reissued when for some valid reason the scope of the original proposal is altered. If you are setting up a new program, guidelines to meet your objectives and operation should be tailored to meet specific needs.

The Meeting Activity

Minutes for each meeting are prepared by the secretary. A copy of the agenda is provided for each individual attending the meeting. A typical format is shown in Figure 7–7.

The approved minutes provide a list of the accomplishments for

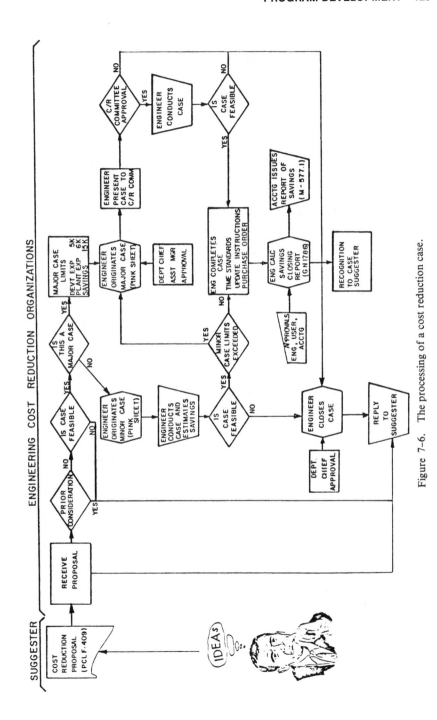

Figure 7-6. The processing of a cost reduction case.

COST REDUCTION

COMMITTEE MEETING MINUTES

Advanced Products, Inc.   Meeting #26

Sunnyvale, California   March 17, 1981

Committee Members	Cost Reduction Coordinators	Guests and Cost Reduction Conductors
F. Wilson, Chairman	D. Quick	E. Peck, Manager
D. Franks, Secretary	P. Hawk	W. Matt, Engineer
P. Cotter	M. King	R. Mech, Engineer
E. Jones	B. Walker	J. Criner, Engineer
A. Lind		

1. The Cost Reduction committee met on March 3, 1981, at 10:00 A.M. in the Executive Dining Room.
2. The following closing reports on completed cases were reviewed and approved. ($ in 000)

Case No.	Description	5-Yr. Avg.	Current	Return
a. 245-A	Reduce Use of Precious Metal F. R. Davis, Engineer	$ 60.5	$ 53.7	190%
b. 186-C	Replace Manual Bonders R. T. Mace, Senior Engineer	$235.1	$195.6	215%
c. 205-A	Redesign 297 Spring Assem. A. C. Jones, Engineer	0	0	0
d. 169-B	Revise Insp. Plan for Wafers E. L. Mann, Engineer	$ 25.7	$ 18.5	160%
	CLOSED	$321.3	$267.8	xxxx

3. The following new cases were opened and approved by the committee.
a. Case No.   Scope: To reduce the cost in the application of nickel silver raw material. This case deals with savings of labor and material.

Expenditures			Savings		
Dev. Exp.	Assoc. Exp.	New Plant	5-Yr. Avg.	Curr. Yr.	Return
$10,000	$1,500.	0	$51.8	$50.5	300%

Suggested by   Conducted by

F. H. Wall, Engineer   F. H. Wall, Engineer

b. Case No.   Scope: Replace oversized supply fan motors with smaller energy-efficient
   280-A   motors. New 10-hp and 15-hp motors will replace the existing 20-hp
                      motors.

Figure 7-7. Committee meeting minutes.

Expenditures			Savings		
Dev. Exp.	Assoc. Exp.	New Plant	5-Yr. Avg.	Curr. Yr.	Return
$3,000	$3,000	$22,000	$14.8	$14.8	52%

Suggested by   Conducted by

O. L. Lane, Supervisor   J. W. Gordon, Engineer
J. W. Gordon, Engineer

c. Case No.   Scope: Reduce solder time on the 973-B Sub Assembly. At the present time,
_____   soldering is performed using a Weller 60-watt iron with an 800°
280-C   solder tip. This proposal deals with increasing operator output by
                       using a 60-watt iron with a 900° solder tip.

Expenditures			Savings		
Dev. Exp.	Assoc. Exp.	New Plant	5-Yr. Avg.	Curr. Yr.	Return
$10,000	0	0	$109,200	$110,300	700%

Suggested by   Conducted

J. B. Wilson, Engineer   J. B. Wilson, Engineer
S. D. Wesley, Senior Engineer   S. D. Wesley, Senior Engineer

4. The following cases scheduled for closing were extended with revised closing dates.

Case No. Date Opened	Description	Est. Savings Current Level	Revised Closing	Reason
a. 201-C 7/80	PCB Assem. Methods Using Single Station Concept. C. R. Davis, Engineer	$42.6	9/81	Design Change
b. 209-A 9/80	Imp. Quality on Gp. B Modems L-1, L-2 R. L. Mason, Engineer	$36.5	10/81	Program Delay
c. 214-C 12/80	Reduce Assem. Time on the RLX-Modules R. L. Jones, Engineer	$67.3	11/81	Waiting for Machine

5. Unresolved Items - None

Approved _____   _____
               Chairman   Secretary

Copty to:
R. L. Harris - Austin
D. R. Wilkes - Denver
H. D. Krahn - Raleigh

Note: The cost reduction suffixes A,B,C, denote product class of business.

Figure 7-7.   (*continued*)

each meeting. If your program operates in a multiplant environment, copies can be forwarded to each location. In most cases this can be done best through the efforts of the cost reduction coordinator. The structure that you operate in may dictate a different distribution pattern. The means by which the data is communicated is a secondary matter. The important aspect is that the data is shared with others in the organization regardless of how the organizational chart looks on paper.

Each meeting provides an opportunity to update progress that has been achieved to that point in time. Refer to Figure 7-8 for a typical example. A review of the first three months' achievement is shown for each product group. The column previously closed is a statement for January combined. Closed this month is a statement of achievement for March. The yearly total is summary for the first three months. The adjacent column labeled "yearly goal" is the planned achievement for the year. The percent complete column indicates the present achievement as compared to the yearly goal. Second-quarter goals are indicated as a projected statement. The third- and fourth-quarter goals are stated in terms of the balance that must be achieved in that time period. Normal progress against planned goals will change each col-

ADVANCED PRODUCTS, INC.
COST REDUCTION SAVINGS

MARCH 25, 1981

PRODUCT CLASS	PREV. CLOSED	CLOSED THIS MO.	YEARLY TOTAL	YEARLY GOAL	% COMP.	GOALS TO 2ND QTR.	COMPLETE 3RD & 4TH QTR.
A	$ 29.7	$ 53.7	$ 83.4	$ 425.0	19.6	$ 55.0	$286.6
B	$ 43.1	$ 18.5	$ 61.6	$ 200.0	30.8	$ 87.6	$ 50.8
C	$ 32.7	$195.6	$228.3	$ 750.0	30.4	$165.3	$356.4
D	$ 21.9	$ 0	$ 21.9	$ 100.0	21.9	$ 42.7	$ 35.4
E	$131.3	$ 0	$131.3	$ 300.0	43.7	$102.4	$ 66.3
F	$ 62.7	$ 0	$ 62.7	$ 350.0	17.9	$ 85.8	$201.5
TOTALS	$321.4	$267.8	$589.2	$2,125.0	27.7	$538.8	$997.0

Chairman

Figure 7-8. Cost reduction status report.

umn monthly with the exception of the yearly goal. There are occasions where the yearly goal may change. These changes will be inadequate without major changes in the planned level of business.

RESTATED COST REDUCTION OBJECTIVES

After reviewing the monthly meeting of the cost reduction committee,
it appears appropriate to restate the goals and objectives:

- To stimulate cost reduction ideas and successful follow-through
  in all phases of the business. This includes clerical operations as
  well as the ever-present challenges in the production and distribution network.
- To provide a medium for exchange of new ideas and methods.
  The development and sharing of ideas is the basis of any successful program.
- To reduce duplication in the pursuit and development of cost reduction. This aspect of the program can be enhanced by effective
  coordination.
- To provide management recognition for excellence by individuals
  and other groups in the business structure. Included in this
  grouping are engineers, managers, and nonmanagement employees.
- To measure yearly performance obtained as measured against
  yearly goals. The yearly goals may be stated in several different
  ways. The most common may be, "Our goal is $6 million in
  1982." Another approach may be to state the savings goal as a
  percent of total sales for the period.

SUMMARY

The intent of this chapter has been to discuss development concepts, the function of the cost reduction meeting, the roles that
team members play, and formats that can be used to document as
well as process ideas.

Cost reduction summary statements can be designed to fit your specific needs. Your program will be, as most programs have been,
changed as your needs and business conditions dictate.

8
Expense Control Techniques

Development of expense control techniques is an ongoing challenge for management, engineers, and other members of the work force. This challenge is not viewed as a temporary issue. In fact, the opposite may be true: The reduction in expenses will continue to be a goal in every company.

In general, reduced expenses are usually linked to increased productivity. There has been a great deal of controversy in the past few years about the low growth rate of productivity in the United States. The slowdown has left its impact on the total economy. Many discussions have focused on *macro* developments that have led to the slowdown in the economy.

Actually, productivity gains are largely made on a *micro* basis. They are made by individual firms and are made on an item-by-item, process-by-process, and plant-by-plant basis. In summary, gains made in these areas contribute to the total productivity at the firm level. In order to achieve increased productivity at the firm level we must be able to control and reduce our expense inputs.[1]

THE CHALLENGE

In order to do this, we must be able to analyze our current productivity measurements; from this we must be able to determine our weak spots. To do this we can start by looking at three areas of expense.

- Direct labor.
- Indirect labor.
- Related expenses.

Take a good look at each of the three areas. If things look good in each group you could be looking in the wrong place, or perhaps you just have not looked far enough.

Ask yourself the following questions concerning each expense grouping.

Direct Labor
- Do proper man-loading techniques exist?
- How often is man loading validated?
- How is excess man loading determined?
- Who makes the initial and final decision?

Indirect Labor
- How do you control and pay for absences?
- How much training is really needed?
- How do you control miscellaneous expenses?
- How do you control travel and living expenses?

Expense
- How do you control expense supplies?
- How do you control janitorial service contracts?
- How do you evaluate the required guard services?
- How do you control energy costs?

Your first thought may be that the above items will all take care of themselves. This is not so. These items are in fact the basics of any cost reduction program. Items such as these have provided stepping-stones for many successful cost reduction programs. What is the secret in gaining control of the above items? The secret is the involvement of people. In this case the ideas can come from people in all strata of the corporate society.

THE MAZDA EXPERIENCE

The Mazda Company has developed a close-working management-labor relationship that has resulted in a strong employee interest in all aspects of company operations. Management encourages this atmosphere of involvement and has developed several systems that deeply involve both labor and management. An employee suggestion program has operated successfully for more than 25 years. Any company employee can suggest an idea for improving operations in the work environment. This encouragement policy in 1981 led employees to submit more than 1.75 million suggestions. The acceptance rate for all employee suggestions exceeded 50 percent.

Quality circles provide another valuable forum for employee participation. In 1981 there were approximately 2150 circle groups. Over 16,000 employees engaged in ways to solve problems in an effective manner. The problems solved were related to quality, productivity, material consumption, and other production-related items. A quality circle will usually consist of seven or eight members. The members meet on a volunteer basis. They meet on their own time several times each month to discuss and solve issues of mutual concern. These groups have made contributions to increased productivity and quality. Aside from this, the quality circles have provided the employees an opportunity to participate and gain a sense of accomplishment.

The question to be answered at this point is, does such a system really work?

A valid question indeed. Let us look at the results that have been achieved.

- Mazda productivity has increased 97.8 percent in the five-year period of 1976–1981.
- Employee suggestions made and accepted have grown at a steady rate. Approximately 200,000 suggestions were submitted in 1975 compared with approximately 1.75 million in 1981. The acceptance rate has been above 50 percent.
- Translated into a partial productivity measurement, 24 vehicles were produced per employee in 1976 compared to 46 vehicles in 1981.
- The bottom line in 1981 was an after-tax income of $85.0 million. This achievement was based on 1.25 million sales units in fiscal 1981.[2]

This summary speaks for the efforts of a group of employees who are involved with and care for the company that provides them the weekly paycheck and a secure way of life. Statements like this are not unusual. Examine the annual report of your company. Look for statements that indicate current performance in the area of cost reduction.

CLOSER TO HOME

In June 1979, the Westinghouse Marine Division introduced a divisionwide productivity improvement program. The program was introduced as a P.R.I.D.E. program. The objective of this program is to identify effective methods for generating ideas, develop ways to process and handle the ideas, and implement programs that reduce costs. The end objective is to increase productivity. This venture was a stepping-stone to the formation of an effective quality circles program. The program was developed along the following lines:

- Quality circles are small groups of volunteer employees that meet for one hour each week on company time.
- Group size would run from 6 to 10 people doing related work assignments. The objective of the meetings is to identify, analyze, and solve work-related problems.

Quality circles utilize a set of techniques that involve individuals who work together in the problem-solving process. Because no work environment is perfect, the people who face the specific problems come together to identify, investigate, and find solutions to those problems. Problems do not just happen—they are created. Differing goals, work schedules, tasks, and changes can all create problems. Some problems are caused by people, some problems are caused by materials, or methods, or machinery, and some very complex problems can involve people, materials, methods, and machinery. Identifying the problem is not the end of the challenge. Quality circles go on to investigate exactly what caused the problem, determine how best to solve the problem, and put that solution into practice.

The quality circles program has a step-by-step procedure for solving problems called the quality circles cycle. At each step a question is answered or task completed.

- What problem are you going to attack and why?
- Is this one problem or several problems in one general area? Simplify by choosing one problem and research that area.
- Sort the information from your research. What factors are a part of the problem?
- What steps are you going to take to solve the problem?
- How do you implement those actions?

- Put your implementation plan into action.
- Have you really solved the problem?
- Have you set up a way to make sure that the problem stays solved or that your solution does not cause a different problem?
- Are there any areas of your problem left unsolved? Investigate and solve those areas.

A management review will take place when the quality circle is ready to share its progress. In some cases the group may solve a number of related problems before the review with management. Some types of problems will require that management become an active force in the various stages that develop the solution. In either case the management review permits an opportunity for the group to share its progress.

How has the system worked? From the concept development stage in 1979 to mid-1982, over 1000 quality circle groups have been formed within the Westinghouse Company. Feedback from several close friends of mine indicate that the program is viewed as a success in every sense of the word.

OTHER APPROACHES

When you view the various facets of your own job, or one you are familiar with, can you find ways to improve functions or save money or time? If you can, then you have already completed the first step of developing a cost reduction. Take just a moment and make notes on some idea to follow up at a later date. Cost reduction—whether it is in the area of engineering cost reduction or the better-known employees' suggestion system—has been lowering the costs in many companies for more than 30 years. And now, with today's environment of high inflation, soaring interest rates, and increased material and labor costs, it has become even more vital to review current practices and material to determine if there is a better way.

"Think, Write, Submit" is a new action motto for cost reduction to encourage employee participation in the engineering cost reduction system and the suggestion system. This is essential for continued success in the program. It is important that this thought, action, and follow-through chain not be broken. If you do not write it down now and submit it, someone else may do just that. This has happened to me twice in the last year. Perhaps you have experienced the same thing. Timing is a key element for successful involvement. The second step in the process is writing down the idea on a proposal form including the present method, your proposed method, and areas where savings might be realized.

The Proposal

After the proposal is submitted into a typical follow-up system, it is dated, entered into the tracking system, and assigned to the proper investigator. The investigator performs an economic feasibility analysis and collects supporting data for the proposal. At this point, the proposal may become implemented as an actual case. After implementation, and after the case is closed, savings are documented.

Engineering cases can focus on reducing labor and material cost through technological innovations, utilizing space requirements better, and reducing energy costs by becoming more energy efficient. As a general policy most companies exclude engineers from direct awards for their efforts in this area since in most cases this is the main reason for their job assignment. However, this is not always true. An exception to this general rule would be an engineer who developed or generated a new method, material, or a new design that was beyond the scope of the normal assignment. In this case the individual could indeed be considered for an award.

Pacific Gas and Electric Company has an employee suggestion system that can award up to $25,000 for an idea. All company employees are eligible to participate voluntarily in the suggestion program. Some matters are not generally processed by the suggestion system. Ideas concerning union negotiations, executive-level decisions, or company policy, or those where the company does not have the power to implement, are excluded. In some cases a suggestion may be submitted jointly in the names of two people. As a general rule not more than two people will be included on the same suggestion. All suggestions become the property of PG&E, and all rights to the idea are assigned. It also follows that the company has the final say concerning adoption, rejection, and awards. One thing is for sure, the $25,000 potential award does generate some unique ideas.

A TYPICAL SUGGESTION PROGRAM

I have examined a number of suggestion programs in use in small companies, large ones, and even very specialized programs designed to fit the needs of a small college campus. In general, many such programs have points that are quite similar but yet are very different in application of the guidelines. Usually one major difference will be in the maximum dollar value of award that is associated with the program. Each company will have special needs. With this thought in mind let me share some guidelines for drafting a suggestion program that can be altered to fit your needs. The key points of such a program will deal with the following:

- Eligibility rules.
- Adoptable subjects.
- Ineligible subjects.
- How to develop suggestions.
- What a suggestion should cover.
- How a suggestion is processed.

Guidelines for Your Program

1. *Eligibility rules.*
   a. Every active employee of the company is eligible to submit suggestions.
   b. When the same suggestion is submitted separately by two employees only the employee who first submitted the suggestion is entitled to an award. Priority is determined by date of receipt by the program administrator.
   c. Suggestions not adopted will be kept in active file for one year after notification to the suggester of the result of initial investigation. Employees remain eligible for an award during this period if changed conditions warrant adoption of the suggestion.
2. *Adoptable subjects.*
   a. All products manufactured; processes or methods used 60 days after their layouts, prints, designs, or instructions have been issued.

b. A suggestion should accomplish at least one or more of the following:

COMBINE   Operations, functions.

CREATE   New products, designs, methods, or services to customers.

ELIMINATE Bottlenecks, duplication, unnecessary operations, clerical work, or reports.

FOOLPROOF Fixtures, machines, procedures, safety.

IMPROVE   Accuracy, quality, tooling, equipment, product design, fixture and tool designs, procedures, systems techniques, storage, packing or shipping, production control, machine performance, material handling, housekeeping, working conditions, layouts, paperwork, shop or office efficiency, security, service to customers, communications.

REDUCE   Rework, scrap, tool breakage, personnel or property hazards, waste, maintenance, repair, downtime, man-hours, costs.

SAVE   Time, space, material, supplies, manpower, utilities.

SIMPLIFY   Designs, procedures, forms.

3. *Ineligible subjects.*
   a. Those which pertain to routine maintenance.
   b. Suggestions dealing with company policy on collective bargaining matters, employee benefits, service anniversary gifts, hours of work, rates of pay, job grading, vacation, and organizational matters.
   c. Proposals already under active consideration by the company.
   d. Ideas previously suggested by another employee during past 12 months.
   e. Suggestions that are part of the suggester's job responsibilities at the time of submission.
   f. Calling attention to typographical errors in drawings, specifications, and handbooks unless they represent discrepancies of long standing and have caused a significantly increased cost to the company.

g. Routine improvements to newly occupied premises during the first six months of use (except for safety or security suggestions).

h. Ideas that one can put into effect without anyone else's approval.

i. Changes that employees are normally expected to submit or develop in the course of their assigned duties.

4. *How to develop suggestions.*
   a. Concentrate on areas you know best.
   b. Keep abreast of the times; read, observe.
   c. Forget tradition. Turn your imagination loose.
   d. Do not disregard an idea because it seems too simple.
   e. Do not put off turning in your idea. When you get an idea, write it down right away.

5. *What a suggestion should cover.*
   a. A description of equipment involved. Use standard code identification when possible.
   b. All information requested on the face of the suggestion form. Describe your idea as clearly as you can. If you have difficulty in describing your idea, your supervisor will be pleased to assist you, or if you prefer, you may outline your idea in a general way and add a request for assistance. A suggestion investigator will be assigned to help you.
   c. A specific method for achieving a savings or benefit.
   d. Drawings, sketches, photographs, and any other helpful information.
   e. If you refer to a form, a sample of same.
   f. A separate suggestion form for each idea submitted.
   g. Date and signature on your suggestion form. When two or more employees submit a suggestion, each must sign.

6. *How a suggestion is processed.*
   a. When your suggestion is received, it is dated, numbered, recorded, and acknowledged.
   b. Investigation is started as soon as possible by the organization concerned with the subject.
   c. You will be notified as soon as the investigation is completed, or of investigation's progress within 30 days of the suggestion's receipt and at 30-day intervals thereafter if the investigation continues into such periods.
   d. If you are eligible and your suggestion is adopted for use, you

will be granted a cash award. Cash awards range from a minimum of $25 to a maximum of $25,000 for suggestions that produce tangible savings. Awards for suggestions that produce intangible savings are cash awards of $25–$100. Awards are based on estimated savings to the organization to which the suggestion applies. A supplemental award will be paid if your suggestion is applicable to, and put into use by, other company locations.

   e. If your suggestion is not accepted you will receive a letter explaining the reasons for this decision.

   f. Your suggestion will be protected for one year from the date of the declination letter. If you request a reinvestigation of your suggestion before the expiration of the protection period, and the suggestion is still not adopted, the protection period will be extended for another year.

A Wrap-Up

The first step—and perhaps the most vital step of the program—is arriving at the actual suggestion. The suggestion can be related to the improvement of methods, products, equipment, procedure, work conditions, safety, or reduction of time or expense. Maybe the idea involves eliminating a bottleneck in a time-consuming procedure or reducing downtime of machinery. Through the employee suggestion program, these and other ideas submitted by employees are investigated to determine their feasibility.

A suggestion must include a specific method for achieving a savings or an improvement, not just a statement that something should be done. A couple of examples that may stimulate some related thoughts are:

• I suggest that a new system to control and measure the use of expense supplies be introduced. How it will work: Each department will be charged with all expense items that are ordered. Also, a review of all stocked items should be completed to determine which ones should be discontinued. First-year savings from these two actions could be an estimated $25,000.

• I suggest that Department A and Department B be combined under one manager. How this will be done: Department A contains 12 engineers and Department B contains 14 engineers. Both groups provide an engineering service for field operations. Savings from this move will include the salary of one department chief and one secretary. Estimated savings $65,000 in the first year.

Planning is the key word that is associated with any successful cost reduction program. A successful program will be linked with the following planning trademarks:

- A well-defined organization.
- A clearly defined charter.
- A clear statement of goals and objectives.
- A month-by-month measurement plan.
- A feedback system to all employees.
- A recognition system for outstanding performance.

There is no doubt about it, the planning or lack of it will be apparent in the results that are obtained in any ongoing program. With this thought in mind, the following concepts by Dr. Frank E. Cotton, Jr., illustrate the outcome of successful planning. The paper deals with a wide range of planning applications. These concepts can be applied to various problems that confront today's managers and engineers. Regardless of your present assignment or past achievements, a careful review of these guidelines should be beneficial.

> *Productivity doesn't just happen by "trying harder."*
> *It must be planned.*
> *But how do you plan for productivity,*
> *and what factors are involved?*

> Frank E. Cotton, Jr.
> Mississippi State University

IN PRODUCTIVITY, "PLANNING IS EVERYTHING"*

Although productivity improvement is by no means new to private business and industry, there now is a new surge of interest. Hopefully, through effective planning, this surge will evolve into a continuing, intensive development and will permeate the public sector of the economy as well.

---

* Reprinted with permission from *Productivity: A Series from Industrial Engineering.* Copyright © American Institute of Industrial Engineers, Inc., 25 Technology Park/Atlanta, Norcross, Georgia 30092.

Certain Assumptions

As a current basis for examining productivity planning, three premises are offered. These describe the setting and the philosophy for this examination.

1. Planned change is essential. We live in a highly dynamic era, with competition, population expansion, depleting resources, pollution, social demands, and many other pressures dictating change. Furthermore, all improvement involves change, but all change does not result in improvement. Thus, change, to cope with these needs and to assure improvement, must be planned change.

2. Productivity is concerned with both efficiency and effectiveness. Usually productivity is defined as being the same as efficiency: a ratio of output and input. But one might ask, "productivity for what?" or "output of what?" Effectiveness is the extent to which the objectives are reached or needs are met. It is appropriate that the concept of productivity be concerned with output not just as a quantity of goods and services, but with output in its relationship to the utility or value it ultimately provides. Often we can provide better end results with less output and less input. Consequently, our output focus and measures should be based on the ultimate results we are trying to achieve. Productivity, therefore, is concerned with how well we achieve our objectives and the total resources required to achieve them.

3. Productivity must be an integrated and continuous part of all functions. Productivity should be concerned with every function, each act and activity, and all resources. Each objective of an organization and its components should be pursued effectively and efficiently. Thus, the pursuit of productivity should be an inherent part of every responsibility. This does not avoid the need for special functions directed to specific productive improvements or providing technical advice and assistance in achieving productivity. Just as change is continuous and permeates all that we do, so the productive pursuit of change must be a continuing responsibility of every function.

Productivity Through Planning

The very purpose of planning is to obtain better results, to more effectively and efficiently achieve our goals. This, of course, is what productivity is all about. Planning, itself, can be highly productive, as it increases the productivity of other functions and activities.

Henry Ford once said that if you need a piece of equipment but don't buy it, you pay for it even though you don't have it. The same is the case with planning: You either invest in planning and improve your results or you pay for not planning through the lack of results. We will examine the planning process and how it can be most productive.

The plan is nothing. Planning is everything.

General Dwight D. Eisenhower

Planning is not just for staff planners; it is for management at all levels as well as supportive and advisory staff personnel. Planning should be a part of every job and should be a basis for decisions and activities. Generally, the higher up the organizational ladder one ascends, the larger the portion of his time should be allocated to planning. Nevertheless, planning can be highly productive at all levels, and the lack of it usually results in failure or less-than-optimum performance.

Planning, as a goal-oriented process, should have three stages:

1. Preparation.
2. Planning.
3. Performance.

A model of these three stages and the supportive information system is shown in Figure 8A–1.

"The preparation stage" provides the basis for planning and effective long-term performance. Here, principal goals and barriers to their achievement are determined. Human relationships are developed, and tentative alignments of key people are structured. Planning procedure is formulated to assure that the whole process of planning will be both effective and efficient. The omission of this stage, as often occurs, is analogous to building a structure without a foundation.

"The planning stage" provides evaluation and creative adaptation of the plan to reality. The planning activity of this stage determines the fate of the plan: whether it is regenerative, thus correcting and strengthening itself, or is degenerative, diminishing its results and its ability to survive.

The planning model, Figure 8A–1, encompasses an information system, without which all three stages of the planning process are jeopardized. This is due to three strategic functions:

1. Input of information relevant to the information of plans;
2. Information interchange by which the various departments share information during performance; and
3. Information collection for corrective action and improvement, Reference.

Although the model shows a downward movement through the planning process, there is feedback among these stages through the information system. This feedback is essential if the process is to be productive. Thus, we

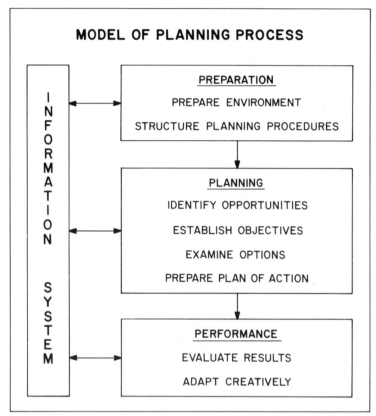

Figure 8A-1.   Model of planning process.

have a type of corrective action, or adaptive improvement, that enables planning to strive toward better and better results.

The relationship between planning and productivity is even more apparent from further examination of the four major components of the planning process.

Stage I. Preparation: Planning must be planned, especially if it is to be productive. Inadequate preparation for planning results in uncertainties, lack of support and follow-through, and minimal participation and benefit from various insights. It may even cause direct or subtle sabotage of activities being planned. How many cost-reduction projects, new equipment installations, personnel reorganizations and employee payment plans have been ineffective or excessive in cost because the preparation stage was omitted or insufficient?

Detailed planning procedure, tailored to the specific need, not only expedites planning but also provides the basis for continued improvement of each planning procedure. Procedures for planning changes in work force, alterations in a production process, new programs, style changes, and long-range projections, and many more cyclical activities, should be used and reused to make planning both effective and efficient.

Unfortunately, many of the adverse repercussions of inadequate preparation are long-term and difficult to establish as a cause-effect relationship. A new process for assembly may provide an apparently large cost reduction, but it may gradually become inefficient as employees fail to give it their support. Or a new pay plan may provide excellent initial response of employees but later become costly and ineffective because the supervisors were not involved in the planning. These are symptoms of improper preparation; the impact may be obscure, but it can have a major negative influence on productivity.

Stage II. Planning: Each element of this stage contributes to productivity. Opportunities are identified and evaluated against specific objectives. Evaluation of options explores the degree to which the opportunities can be exploited, and it examines the barriers that could arise and the productivity likely to be achieved. When the options are selected, a plan of action is prepared to assure that development occurs with high effectiveness and efficiency.

This subprocess not only provides a plan that is more likely to be highly productive; it also involves the concerned people in the process and thus benefits from their thinking and aligns and motivates them to work toward achievement of the objectives they helped to establish. Execution of a well-planned plan is less likely to be confronted with major, unexpected problems, and the chances are greater that contingencies are provided for so that target objectives are achieved.

Stage III. Performance: The very purpose of the performance stage is to assure productivity—that objectives are reached effectively and efficiently. It is through these efforts that deviations from a target course are detected and corrected. Here, continuous adaptation of the plan to reality is made, which means careful surveillance and reactive remodeling of the plan throughout its implementation.

The performance stage essentially provides a detection of low productivity and a feedback of information to appropriate points to assist in reaching the objectives productively.

Information system: Productive planning relies on the right information at the right place at the right time. This implies not only an efficient selection and collection of the pertinent information, but also objective analysis of it and its prompt distribution or availability.

Since many forms of planning in an organization must be periodic or continuous, so must the flow of much information. In addition, further information often is needed for special studies as a basis for extraordinary planning.

The productivity of all planning is sensitive to the information system. The data system and the people-to-people communications comprise an information base on which planning must rely. Even the organization structure of the people influences the need for information and the degree to which it is available. Thus, if planning is a productive agent, the information system must be reliable and effective for this productivity to materialize.

Each of the four components of the planning process has a distinct contribution to productivity. Yet, it is the whole process that makes possible high productivity results. The omission of any element of the process can seriously reduce that potential.

To assure that planning is effective, however, the whole process must be highly participative. It is the proper involvement of people that converts the mechanics of planning into a living, productive process. General Dwight D. Eisenhower said, "The plan is nothing. Planning is everything."

Effective planning is a process of human interaction, involvement in the planning of those who will be concerned with its implementation, and formulation of objectives by those who will pursue them. This human element must permeate the whole planning process if those who are to carry out a plan are to feel a personal identification with it. Otherwise, they are unlikely to be enthusiastic and dedicated to assuring its success.

It is during the planning process that ideas must be tapped, barriers and misgivings must be identified, and conflicting interests must be worked out. It is during the planning process that the future of the plan is molded.

How Planning Improves Productivity

1. Planning identifies future productivity possibilities and prepares for their possible adoption at the most opportune time.
2. Planning is a means of molding the future rather than merely reacting to events as they arise. To assure that change is productive, it requires careful planning.
3. Effective planning encompasses the use of objectives to focus action, motivate personnel, and evaluate achievement. These are essential to productivity.
4. Planning stimulates creative thinking, reduces fear of the future and resistance to change, evolves productive teams, and provides an effective means for their collective reinforcement in striving for difficult goals.
5. The planning process provides the most effective medium for line-staff relationships and for vertical and horizontal cooperation in an organization.

Planning for Productivity

Productivity must be planned. But this planning must be pursued by all management and staff personnel, and it must permeate and affect all that is done—by everyone—in the organization. Consequently, planning for productivity is a decentralized but integrated effort to make all planning, decisions, and activities as effective and efficient as possible.

Productivity planning, as a successful, long-term undertaking involves three principal steps:

1. Developing an effective planning process and structure in the organization.
2. Preparation of productivity goals, and permeating the planning process with specific objectives based on those goals.
3. Establishing productivity surveillance, assistance, and coordination in a manner tailored to the organization's needs.

The planning process, Step 1, was presented in the previous section. The planning structure, the organizational relationships through which the planning process occurs, must be tailored to fit the total organization structure. If planning is to be maintained on a highly successful level over a long period, it must be a normal function of management. Planning assistance, data analysis, and special studies can be performed by staff personnel; but if management does not sufficiently engage itself in the process of planning, their decisions and the results of those decisions are likely to be inadequate. A sound planning process and structure is a medium through which productivity is developed.

Productivity goals (Step 2), developed by top management, are based on such factors as competitive pressure, realistic potential for productivity achievement, conflicting goals, external requirements and constraints, and management policies. These goals provide the basis for setting specific productivity objectives by each unit of the organization. The objectives, then, with target levels and dates and assigned responsibility for their achievement, become integrated into the cyclical process of planning, execution, and control. In short, productivity becomes a normal and continuing part of the function of every manager and every unit.

Step 3 is a facilitator in the productivity planning and development. It is necessary as part of the information system of the planning process, but it also provides direct assistance to management by supplementing their limited time with staff expertise. This type of staff assistance often is provided by industrial engineers, systems analysts, and planning specialists. Their role is to assist and reinforce management in the difficult job of planning, not to do the planning for them.

The surveillance and coordination of productivity planning is a management function but needs technical assistance somewhat equivalent to financial planning and control through the comptroller's function. Some of the tasks that staff assistance can provide to management include developing standards and measurements of productivity, gaging the corporate productivity progress relative to its competition, analysis of external influences on productivity and how they can be improved, and surveillance of the whole process of productivity planning.

Productivity Factors

Special Programs. Many types of programs are used for boosting productivity. These include such well-known concepts as suggestion systems, incentive plans, value analysis, zero defects, management by objectives, and participative management. Many such programs have much to offer, and each organization should carefully study their own needs and situation, then select and mold special programs into their system in a sequence that is reinforcing rather than taxing. Such programs should be individually explored in advance to assure their long-term survival and potential benefit.

Some programs tend to be short-term assets but long-term burdens. Their demise, however, may poison the setting such that new or replacement programs are resisted. Dr. Harold Enarson, President of Ohio State University, observed that, "Too often our planning efforts display an obsessive preoccupation with whatever is fashionable" (2). He noted that this human tendency applies to techniques, tools, and programs. Fads are difficult to resist, even if they may be costly.

> *Project timing is critical. A new project should allow a settling down of prior projects, but begin before any decadence occurs. . . .*
>
> Dr. Frank E. Cotton, Jr.

Implementing Planning. One of the most formidable barriers to planning, including productivity planning, is how to get people to plan, and plan effectively. We resist planning because it requires conceptual thinking and projection into the future. It is easier for most people to think in terms of things rather than concepts and to think about what is happening now than about what might happen in the future. There is a sort of fear of the unknown and the uncertain, or at least a fear that one may not be able to cope with what may transpire—a fear of failure.

How, then, can a transition to effective planning be made? This transition must be planned carefully. It must provide an evolving development of planning skills and self-confidence over a substantial period of time. The steps must be small and properly sequenced so that the unfamiliar is minimized and confidence gained in one step provides clarity, confidence, and readiness for the next.

For three reasons, planning development should be done initially in existing and familiar jobs:

1. The amount of unfamiliarity is reduced by applying new planning techniques to problems that are routine and familiar.
2. Confidence is gained with planning methods and conceptual projections before trying them out in new projects; and
3. Competition for scarce time needed for immediate matters is minimized by first applying planning to current, pressing problems. As planning momentum evolves through increased confidence and success, venturing into new techniques and new projects becomes a natural result.

Productivity planning in no different, even if a sound process and structure of overall planning already exists. The early developments should be planned to attain productivity improvement in familiar ways in existing activities, and then, later, expand into new techniques and programs. Competence and confidence are molded through focusing early productivity planning on short, visible, urgent, and easily achievable activity. Through these successes, management and supportive personnel are more motivated and better equipped to undertake more diverse projects.

Productivity Applications. Productivity planning should allocate improvement effort based on the anticipated returns from the costs (including effort costs). The Pareto concept of distinguishing the "vital few" from the "trivial many" can focus attention where improvement can have the largest impact. Bottleneck areas, activities that are highly repetitive, procedures that are highly repetitive, procedures that are used over and over, items that are used in large numbers or quantities, large or repetitive costs, and other high-payoff applications are appropriate for early and continued attention.

Production Techniques. There is a host of available techniques that are used, and often misused, to attain productivity achievements. Some, such as value analysis and a large variety of incentive techniques, were mentioned earlier. Others, such as managerial economics, engineering economy, optimization techniques, and input-output analysis, are available. Significantly, productivity planning should focus on detecting the opportunities for improvement and then applying whatever techniques can provide the best results.

Process Productivity.   Too often, attention is focused on employee productivity; this is a limited view and yields limited results. Productivity attention should focus on the process rather than on just the people who comprise only one of the cost inputs in the process. These processes include not only each of the production processes, but also the technical design processes, purchasing processes, management processes, and many others. Examining the process as a whole often exposes new productivity possibilities. Furthermore, productivity gains accumulate from repetition of productive processes; it is the whole process that cycles, not just the human input.

Information Productivity.   The productivity input from improvement in information and communications is often underestimated. These are highly repetitive processes that have a multiplier influence on many other activities and processes. Misinformation, inadequate information, lack of clarity and relevance, vague responsibility or goals, confused instructions, and misunderstandings—these are some of the information barriers to productivity. The information system and the planning process deserve special attention to assure that these hurdles are minimal.

Productivity Maintenance.   Productivity programs and projects tend to experience creeping decadence or offsetting costs unless maintenance is built into the productivity planning system.

> *If you need a piece of equipment but*
> *don't buy it, you pay for it even though*
> *you don't have it.*
>
> Henry Ford

The productivity growth curve, Figure 8A-2, shows little results during the early period of development, learning and debugging. Then, advancement is rapid, followed by a sloping off and saturation as installation is complete. Without proper maintenance, however, many types of improvements tend to slip away and either disappear, settle to a lower level, or be offset by other costs.

There are sound reasons why this often occurs. It is easier to be motivated while involved in a new development, but when the development is over, the surge of motivation dwindles. Also, some types of support fade away to other projects; many people focus most of their attention to those activities that are under the closest current attention by their management. In addi-

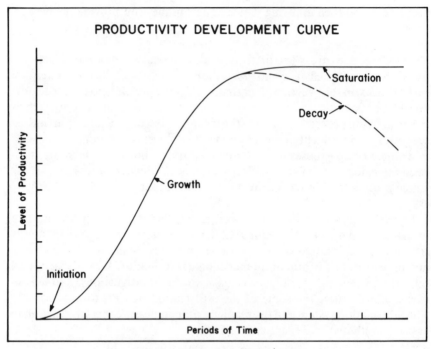

Figure 8A-2.   Productivity development curve.

tion, after a project or program is no longer in the limelight, it is easier for those forces that have been thwarted or ignored to build up increasing pressures that erode the improvement or nullify its benefits. Finally, creeping change usually takes over after a system is installed, and such changes, which are usually negative, often are subtle and inconspicuous, but quite deadly.

Another effective approach to productivity cumulation is to proceed through an endless series of productivity projects. The projects are sequenced so that each one reinforces the prior ones directly or by maintaining motivation and support to the whole system of productivity activity. The timing of initiating each project is critical. A new project should allow a settling down of prior projects, but begin before any decadence occurs or people become firmly settled into new routines.

Productivity Scope

Productivity is rarely, if ever, a primary objective of an organization. It is, however, a major means through which many of the primary objectives are reached. If properly structured, the pursuit of productivity is the pursuit of all the goals of the organization.

Too often, we look for a simple ratio with which productivity can be measured and achievement of objectives can be evaluated. Productivity planning should be focused primarily at achieving goals and objectives. This often means making improvements that cannot be measured directly and in absolute terms. The important results may be indirect and relative. For example, there is no absolute criterion for measuring employee attitude. But a shift in attitude can be detected and expressed in relative terms; such a shift can have significant impact on employee turnover, absenteeism, defect generation, and other factors which can be measured.

Productivity is basically a relationship between output and input. But this is far more complex than the misleading and misused ratio of physical output per employee or units per machine hour. It is a matter of examining all of the outputs, or results, of organized activity and comparing them with all of the inputs, or resources, that are acquired or used to obtain those outputs.

Inputs certainly include all of the tangible resources required for the operations, including management, supportive, and operative personnel; facilities and equipment; materials, supplies, and services; and operating capital. But many more inputs influence ultimate productivity and attainment of objectives. These include such value inputs as community support, free or subsidized facilities and utilities, special tax freedom or treatment, monopoly privilege, restraints on competition through tariffs and import quotas, and government grants and subsidies. Input also includes employee training and education, legal services for advice and litigation, employee services for preparing government reports, and contributions of resources for public or private use.

Too often our planning efforts display
an obsessive preoccupation with whatever
is fashionable.

Dr. Harold Enarson

The array of outputs is even more complex, yet equally related to productivity. Output includes more than products and services. It includes such results as pollution, depletion of mineral resources, employee health and safety, employee development, employee benefits, community good will, community stability, taxes at all levels, and local and national attitudes. Furthermore, when we examine the wider scope of outputs, we become aware of delayed impact of many inputs and activities. Community and national shifts in work attitudes, impact of accumulation of pollutants, and resource availability due to depleting supplies—these are only examples of delayed visibility of some outputs.

This broader concept of productivity and its input and output components enables us to expose some important productivity factors. There are many, but we will note only a few. First, since pollution is an output, it is a definite component of productivity. Input resources to lower pollution levels may not add to the product value output, but the reduction in pollution is a positive change in total productivity. An undesirable (negative) output has been reduced, which is the same as an increase in a desirable output. A "goal" or "utility" has been generated for the public, and resources have been expended to achieve it.

Similarly, increase in the level of occupational safety and health or decrease in potentially harmful exposures is an output. Yet, this output is not in the form of salable products or services, and often causes a reduction in the output of physical products. Consequently, this is an allocation of input resources whose purpose is an output of reduced hazards of accident and health. Traditional measures would record a decline in productivity due to increased input of resources associated with unchanged or reduced output of salable products or services. Yet, from the broader concept, it may have increased due to the increase in a utility for employees.

A third example is the application of affirmative action in the employment of personnel. Traditional measures would indicate a reduction in productivity relative to the degree of resources allocated for training and development and relative to reduced output of products and services due to reduced skill levels, higher turnover rates, and other factors. Affirmative action is a general type of output, a public service, which, based on any deviations from the market economy of employment, may involve additional costs as inputs and reduced output of salable goods and services. The net productivity of these extra resources is based on the public service output.

There are many such outputs that are imposed or voluntarily assumed. A few other examples that are well known are the new pension requirements, government reports required of organizations, welfare programs, social benefits of all types, commodity price supports, and subsidized housing and medical care.

Productivity is a complex concept. If we speak of it in a limited context, we must recognize that there are many inputs and outputs that are being ignored. A broader scope of productivity helps us to examine the complete outputs and for whose benefits these outputs are being produced.

References

1. Cotton, Frank E., Jr., "Bridging the Planning Gap," *Mississippi Business Review,* May 1976.
2. Enarson, Harold L., "The Art of Planning . . . Fugitive Notes of Thinking Ahead," address at The Society for College and University Planning Conference, July 1975.

CASE STUDY EXAMPLE 2

Case Development	Measuring Productivity
Company	United Airlines Chicago, Illinois
Data Source	American Productivity Center 123 North Post Oak Lane Houston, Texas 77024
Estimated Savings	Not Stated

In order to measure productivity within any segment of the industry, an understanding of the inputs and outputs is essential. Consider for a moment that you are part of the ever-expanding airlines industry. Probably your first concerns are to think about the factors that make this industry segment a challenge for engineers and managers.

Measuring Productivity at United Airlines

Introduction.   The 1960s were a period of exceptional growth for United Airlines and the airline industry. Passenger miles were increasing while airlines were enjoying higher productivity due, in part, to the larger and faster jet aircraft. With the introduction of the Boeing 747 in 1969 and the Douglas DC-10 in 1970, the industry anticipated continued productivity gains and higher profits. But the early 1970s saw a downturn in traffic growth, and profits for virtually all carriers deteriorated sharply.

Profile of United Airlines

United Airlines, a subsidiary of UAL, Inc., is the largest airline in the free world. In 1977, with 22 percent of the domestic airline passenger market, it employed over 49,000 people and utilized 352 aircraft which flew over 34 billion passenger miles and generated $3.0 billion in revenues.

While most industries are either labor-intensive, capital-intensive, or energy-intensive, the airline industry is all three. Among United's employees are pilots, mechanics, flight attendants, reservationists, cooks in the food-service kitchens, baggage handlers, and counter personnel—to name a few of the disciplines. These employees are specialized and most are unionized. In addition, there are more traditional corporate personnel such as accountants, attorneys, planners, analysts, and support and management personnel. Wages and benefits compare very favorably with similar jobs in other industries with the result that almost half of United's expense dollar is absorbed by employee salaries and benefits.

On the capital side, United has estimated that between 1976 and 1990 it will need to spend close to $10 billion to replace obsolete aircraft and accommodate modest growth. In 1978, the price tag for just one Boeing 747 was approximately $45 million. Government noise regulations and fuel-related economics have already prompted United to place a $1.2 billion launching order for 30 next-generation Boeing 767 twin jets.

Almost one of every five expense dollars goes to pay United's fuel bills. The precipitous climb in the cost of petroleum from 1972 to 1978 needs no dramatization. The price of fuel to United has increased close to 300 percent since 1972. And only the Department of Defense buys more petroleum products than United.

United's Response to the Downturn

United was caught in the downturn in the early 1970s and sharply felt the economic squeeze. A substantial decline in earnings contributed to a change in top management, and a new president was brought in from Western International Hotels, the other major UAL subsidiary. Several programs were initiated to achieve significant cost reduction. In addition, the new president, a strong believer in accountability and decentralization, moved to restructure the company and to encourage decision making at the lowest possible level in the organization.

However, implementation of the reorganization led to the recognition that United did not have an adequate management information and performance measurement system. A decision was made to develop a comprehensive, integrated Financial Management System (FMS) which included:

- Revenues.
- Expenses.

- Fuel.
- Manpower.
- Productivity.
- Profits.

The relationship among these elements in any company is obvious. While profits may be the bottom line of corporate performance, the other factors are an essential part of management's analysis and understanding of profitability. While most management information systems address revenues, expenses, and profitability, few have incorporated an integrated productivity measurement component. This discussion focuses on United's use of productivity data in analyzing corporate performance.

Productivity as Part of the Financial Management Systems (FMS)

For many years the company maintained productivity statistics in certain areas, although they dealt primarily with flight personnel. In addition, it had an extensive standards program which projected manpower requirements for most ground operations. For all its detail and sophistication, however, the standards program did not enjoy the confidence of operating management and those who developed the budgets. As the manpower standards lost their emphasis, attention gradually turned to productivity.

The computerized productivity reporting system installed as part of the FMS was designed to replace manual and ad hoc measures in use around the company. Consequently, it reported productivity indicators for all divisions of the company at all levels in the organization including:

- Region.
- Station (airport).
- Cost center.
- Shift.

As used in United's system, productivity indicators are ratios that relate volume output and manpower input where there is a correlation between the two. For example, the total product output of a sales and services division is a function of the passenger and cargo handled which is expressed in terms of equivalent passengers boarded. Therefore, the equivalent passengers boarded (output) divided by the number of employees performing passenger and cargo services (input) is the ratio used to measure the productivity of that function. Similarly, other productivity measurements are employed at the division level. See Table 8-1. United used these indicators in making comparisons of two general types—internal and external.

Internal Comparisons. Each year a profit plan is prepared at the cost center level and, in some cases, by shift. This profit plan encompasses productivity and its constituent volumes and manpower in monthly detail. Productivity data are compared at each level against the plan and against the prior year.

Shortly after monthly performance data become available, United's senior management team meets to review results. Each vice-president in charge of a major division presents a report in writing and in person covering his financial and operating results, including productivity. Productivity results from all divisions are summarized.

Although requiring managers to report on and give an explanation for their performance every month does not ensure that their divisions are at maximum effectiveness, it both heightens awareness of productivity and provides a continuing mechanism for review of productivity trends.

Table 8-1. United Airlines Division Productivity.[a]

DIVISION	VOLUME MEASURE	BETTER/WORSE THAN PLAN (%)			BETTER/WORSE THAN LAST YEAR (%)
		VOLUME	MANPOWER	PRODUCTIVITY	PRODUCTIVITY
Eastern	Equivalent passengers	5	1	4	6
Central	Equivalent passengers	1	2	1	4
Western	Equivalent passengers	3	0	3	6
Inflight services	Block hours	1	3	4	2
Inflight services	Available seat miles	1	3	3	6
Flight operations	Block hours	3	1	2	3
Food services	Passengers out	1	2	3	7
Maintenance operations	Equivalent maintenance units	9	8	1	N/A
Finance	Equivalent passengers	0	2	1	5
Company	Equivalent passengers	0	2	3	7

[a] Illustrative only—actual data have been altered.

External Comparisons.   In many respects, notwithstanding the competition between carriers, the airline industry has been treated like a utility. Its services (routes) are subject to approval by the Civil Aeronautics Board (CAB) as are the prices for these services. As a result of its being regulated, it is required to file reams of data with the CAB each year. These data are extremely detailed to the extent that they include such items as ''the amount of money spent on labor in maintaining the airframe of 747 aircraft used in domestic service'' and ''the number of pounds of priority mail carried between Youngstown and Cleveland.'' The data are financial and operational, are filed quarterly or monthly, and even include the number of lawyers and law clerks working for each carrier.

Once filed with the CAB, the data are public. They are quickly converted to computer input and are accessible to interested researchers through private computer data bases. With the ability to research, analyze, aggregate, and dissect being only a phone call away on a computer terminal, the computerized CAB data base has helped defog the goldfish bowl; and no one looks more carefully at the numbers than the participants themselves, United being a prime example.

As the CAB data become available, United's analysts update the ratios that management considers significant in evaluating the company's performance over the period covered. For example, to test its performance in fuel economy against the competition, United might generate the schedule shown in Table 8-2.

This schedule not only tells management where it stands currently with respect to the industry, it also provides information on the trend of its relative performance. This kind of analysis can also be used as an indicator of labor productivity (e.g., number of employees per function), capital productivity (e.g., utilization of aircraft), and general resource management (e.g., load factors and crew utilization).

Table 8-3 depicts an example of a summary comparison listing all expense items and comparing United's performance with the rest of the industry. This comparison provides an opportunity to identify which operational areas appear the weakest with respect to the rest of the industry.

Comparing Performance with Historical Trends

Although the value of comparing performance within an industry is useful in assessing whether a company is operating efficiently relative

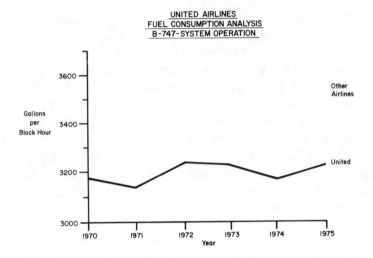

Table 8-2.   United Airlines Fuel Consumption Analysis
B-747-System Operation.

| | GALLONS PER BLOCK HOUR | | | | | |
	1970	1971	1972	1973	1974	1975
SUMMARY						
United	3186	3145	3243	3237	3179	3238
Other airlines (average)	3667	3526	3535	3547	3531	3541
%United better/worse	13.1%	10.8%	8.3%	8.8%	10.0%	8.5%
CARRIER DATA						
United	3186	3145	3243	3237	3179	3238
National	3659	3582	3617	3615	3436	3337
Braniff	—	3336	3408	3356	3341	3356
Northwest	—	3449	3378	3423	3429	3424
American	3510	3416	3381	3384	3353	3429
Pan American	3791	3653	3617	3582	3524	3566
Delta		3420	3529	3562	3560	3633
Trans World	3544	3555	3000	3708	3706	3675
United rank	1	1	1	1	1	1

to other airlines, such comparisons neither reveal the reason for the results nor do they assure the company that it is achieving its maximum
performance. For example, a company that rates high among its competitors may be lulled into thinking there is little room for productivity

Table 8-3.   United Airlines Major Expense Categories. Cost Per Available Ton Mile Ranking.[a]

	% OF UNITED EXPENSE	UNITED RANK IN INDUSTRY [b]					
		1971	1972	1973	1974	1975	1976
Personnel costs							
Flight officers		5	5	6	6	5	5
Flight attendants		5	5	5	6	5	5
Other personnel		2	3	4	6	7	7
Total wages and benefits	47	4	4	5	6	6	5
Fuel	18	2	3	2	2	2	1
Aircraft ownership[c]	9	8	9	10	7	8	9
Aircraft mtce. materials	5	3	2	2	3	4	3
Traffic commissions	4	3	3	3	3	3	3
Food	4	6	8	7	7	6	8
Outside services	4	2	2	1	2	2	2
Facility and eg. ownership	3	5	6	5	5	5	4
Landing fees	2	4	4	3	2	2	3
Communications	1	3	2	2	1	2	2
Taxes	1	3	4	3	2	3	3
Other	2	3	2	2	3	2	2
	100						

[a] Illustrative data—actual rankings have been altered.
[b] Ranked from 1—best to 10—poorest.
[c] Includes depreciation expense for owned aircraft and lease expense for leased aircraft.

improvement. Other measures need to be employed to supplement the results of interfirm as well as intrafirm comparisons.

One of those tests is to compare the company's current-year results with historical data, for example, last year's data or the average productivity for the last five years. This comparison will still not guarantee that the company is performing at its maximum productivity, but it will reveal whether the company overall is continuing to improve at or near a desired rate.

As deviations above or below a historical trend appear, they may indicate a lessening of effectiveness in the particular department or function. However, increases and decreases in productivity can often be explained by seasonal traffic fluctuations, strike effects, revenue yield levels, or even weather conditions. The data must accordingly be adjusted (where possible) for these factors before any firm conclusions are drawn from the productivity indicators.

Future of Productivity Program at United

United's goal over the next several years is to be among the top one-third of the industry in terms of return on investment. Management recognizes that achieving this goal depends on reaching a superior overall level of productivity. Management also recognizes that the competitors who will survive in the postregulatory reform environment over the long run are likely to be those who make the most efficient use of their resources. Consequently, productivity reporting is expected to remain an essential management tool as United pursues its profitability objectives and maintains its leadership position in the airline industry.

REFERENCES

1. Militzer, K. H., Chief Economist, AT&T Co., "Productivity at the Firm Level," Conference on Productivity Research, Houston, 1980.
2. Annual Report, Mazda Company, 1981.

9
Shop Operations—Potential Savings

The modern shop and the not-so-modern shop have one thing in common. Both serve as production centers and generate some type of end product. The product may vary from telephones to very complex microprocessors. The end product may be very different, but a number of elements are common in any situation. These elements can be described as follows:

- Labor input required.
- Material requirements.
- Capital requirements.
- Energy consumption.
- Related expense.

Production jobs have been under close examination for years. Many different approaches have been used to scoop up the savings. In order to be effective in the cost reduction search, a profile developed for your specific operation is a good place to start. Refer to Figures 9-1 through 9-5.

AN OVERVIEW AUDIT

A production audit can be a very rewarding venture in terms of cost reduction savings that can be achieved. Each item shown on the production overview can be dealt with as a separate area of investigation. The approach to the audit can be handled in a number of different ways. The method that is selected will probably be related to your

PRODUCT CLASS A (MISC. ASSEMBLY)   RALEIGH, N. C.

EMPLOYEES DAY NIGHT	WORK POSITIONS STD. MISC.	% 1ST SHIFT CAPACITY	% 2ND SHIFT CAPACITY	COMBINED UTILIZATION
129 20	126 11	94	22	58

SQ. FT. PER EMPLOYEE   TOTAL SQ. FT.----------------------17,914
                                 % OF TOTAL SHOP---------------30.2%
THIS AREA Vs. TOTAL SHOP   % OF TOTAL SHOP HOURS---46.6%
    139   176   % OF TOTAL $ VOLUME-------41.9%

Figure 9-1.   Employee and work position data.

reason for conducting the audit at a specific time on any given profit center. Refer to Figure 9-6.

Three of the most common approaches to the audit are:

• The audit can be conducted jointly by management and engineering. This can be part of an ongoing program that covers selected profit centers on a random basis. The audit, in order to be effective, may cover the entire production overview. If one segment of the production operation is in trouble the "spot audit" can be directed at that element only.

• The quality circles approach may also be an effective method in dealing with selected segments of the operation. One advantage of using this method is the multiple input from the various members of the group with different backgrounds and work assignments. One

PRODUCT CLASS B (CABINET ASSEMBLY)   RALEIGH, N. C.

EMPLOYEES DAY NIGHT	WORK POSITIONS STD. MISC.	% 1ST SHIFT CAPACITY	% 2ND SHIFT CAPACITY	COMBINED UTILIZATION
9 0	9 0	100	0	50

SQ. FT. PER EMPLOYEE   TOTAL SQ. FT.----------------------7,596
                                 % OF TOTAL SHOP--------------12.8%
THIS AREA Vs. TOTAL SHOP   % OF TOTAL SHOP HOURS---3.3%
    844   176   % OF TOTAL $ VOLUME-------3.6%

Figure 9-2.   A contrasting view.

COMPLEX PRODUCTS, INC.

PRESENT SHOP
SPACE ALLOCATION
RALEIGH, N. C.

USE	SQ. FT.	% OF TOTAL
PRODUCTION	59,167	75.1
SHOP STOREROOM	3,534	4.5
AISLES	12,440	15.8
DROP AREAS	1,214	1.6
MAINTENANCE	635	.8
QUALITY	318	.4
RESTROOMS	1,033	1.3
LOCKERS	367	.5
TOTAL	78,708	100%

Figure 9-3.   Present shop space allocation.

disadvantage may be the time interval in developing the audit data since this assignment is only a part-time involvement.

• The services of an outside consultant can be obtained to audit all or any selected part of the production process. An outside consultant can bring a certain amount of expertise to the audit that may not be available inside the company. An outside viewpoint may deliver feedback that someone inside the organization may be hesitant to expose.

COMPLEX PRODUCTS, INC.

RALEIGH, N. C.
SHOP STORAGE MODES

PROCESS MATERIAL IN SHOP	4,075 SQ. FT.
PALLET STORAGE ON FLOOR	2,411
SHELVING W/JUNK MATERIAL	1,241
REPAIRED MATERIAL	1,156
PACKING MATERIAL	931
TOTAL SHELVING UNITS	5,720
TOTAL	15,534

PROJECTED SPACE RECOVERY   4,590

Figure 9-4.   Shop storage modes.

COMPLEX PRODUCTS, INC.

RALEIGH, N. C.

SHORT RANGE PLAN

WHAT	WHY	WHEN
(1) REMOVE EXCESS MATERIAL FROM SHOP (A) PROCESS MATERIAL (B) REMOVE OBVIOUS JUNK MATERIAL (C) REDUCE UNREPAIRED MTL.	TO FREE UP SPACE AND REDUCE INVESTMENT	FEBRUARY 1982
(2) CREATE ADDITIONAL DROP AREAS BY REDUCING SIZE OF THE PRESENT SHOP STOCK ROOM	TO REDUCE CONGESTION IN WAREHOUSE TO IMPROVE MATERIAL/SECURITY	MARCH 1982
(3) REARRANGE MISC. ASSEMBLY REMOVE EXCESS SHELVING UNITS	TO PROVIDE ADDITIONAL WORK SPACE TO IMPROVE THE FLOW OF MATERIAL	MAY 1982

Figure 9-5.   Short-range plan.

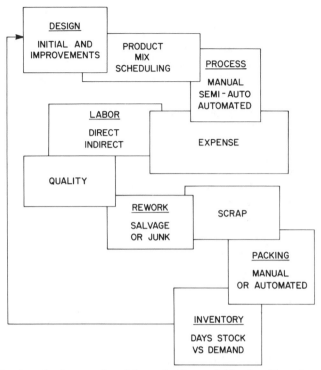

Figure 9-6. A production overview. "Cost reduction can be developed in each segment of the operation. . . ."

The open exchange and total feedback of the audit findings are a real advantage when using an outside consultant.

Your Choice

The approach that is selected for conducting the audit will depend on the company size, financial status, and the mood of management at the particular time. The important thing is to conduct the audit, identify the problem areas, and restructure problem areas into potential for cost reduction. The real problem that can hamper the search for cost reduction is our own attitude and outlook. The potential for cost reduction is in every segment of the operation that we are exposed to on a daily basis. There can be a variety of reasons for this self-imposed immunity that reduces our progress in this area.

A cost reduction audit can in some cases provide a change in our normal approach to the subject. What can be expected as the outcome when the audit is completed? Let us examine some feedback comments that can be the highlights of a typical production audit. The comments are listed under each topic of major interest.

Audit Feedback

According to Webster, as stated in the big book with all the words, the purpose of the audit is to:

- Examine and review a given set of conditions.
- Examine with the intent to insure conformance with a given set of criteria.
- Examine or verify an event to gain knowledge for the purpose of improvement.

An audit of the product design would raise the following questions:

1. Can the final product size or shape be changed to reduce the weight, cube, and shipping configuration?
2. Can special-use parts be replaced with a standard commercial part?
3. Can the package cover, assembly frame, or other related part be changed from a cast or machined part to a less expensive material such as plastic?
4. Can the tolerances—both electrical and mechanical—be adjusted to make the assembly and testing more efficient?

Sometimes even a small change can provide a substantial savings. As an example, consider a piece of electronic test equipment that comes from the factory with a metal carrying and shipping container to insure proper protection and storage. The change to a molded plastic assembly may offer a cost advantage in the range of $4.00 per unit produced. Not a big savings you may say without the real impact of total savings being known. A modest volume of 10,000 units will provide a nominal savings of $40,000 yearly. This is not a potential to be overlooked.

OTHER RELATED ISSUES

Specific questions can be raised about each segment of the operation. The following questions can and should be asked when a production audit is undertaken:

- Can scheduling techniques be improved? How often does the shop change production setups for short-run items?
- Can the process used in the production environment be updated based on the projected product life cycle? If the present assembly mode is manual, can certain steps be converted to a semiautomated or an automated operation?
- Can labor requirements for both direct labor and indirect labor be more clearly defined? Can the job be restructured to reduce the labor grade required? Can the operation be restructured to improve employee job satisfaction?
- Can expense components be controlled in a more effective manner? Can the shop layout be improved to reduce floor space requirements? Can more effective usage be obtained from the energy program? Can internal material flow be improved?
- Can a statistical sampling plan be introduced to replace the present 100 percent inspection? Can the present number of items that drop out at various inspection points be reduced? Can the cost to salvage the item be predicted ahead of time? Can the final scrap rate be reduced? Can the scrap be recycled or sold as a by-product?
- Can the packing operation be improved? Can the special details be replaced with general-use materials? Can carton cutouts reduce the need for duplicate labeling on the carton?
- Can inventory of in-process and completed material be reduced? Can the use of air versus surface shipments be used to reduce inventory?

PROBLEMS AND OPPORTUNITIES

As the profile for your operation is being developed, a number of problem areas will be identified. On the other hand, a number of cost reduction challenges and opportunities will probably develop. One thing that may surface when your profile is being developed could be the lack of a standard cost measurement system. Use of such a system should provide direction in the way the shop is anticipated to move as well as a measure of progress at selected time checkpoints along the way.

In order to secure a firm grasp and achieve maximum cost reduction benefits each product should be analyzed in detail. You may find that some items now being produced should be purchased from an outside source. You may also find that the sequence of operations can be changed to reduce material handling. Other operations may have the potential to be combined or eliminated. Keep in mind that each family of operations is different and unique.

Cost Measures: An Efficiency Yardstick

In any company, regardless of size, it is important to track manufacturing costs to assure that they stay under control. What makes control possible is a standard cost system and its corollary concepts, negative and positive variation. With a standard cost system, any profit center in a company can ask, "How are we doing?" and get a quantitative answer that pinpoints strengths and weaknesses in the operation.

A cost bulletin is a publication of the standard manufactured costs of each of the company's products, large and small, from large systems to small components. A standard cost says, for example, that product A should cost $X to make and a circuit board for product B should cost $Y to make. If a shop incurs greater cost than the bulletin cost, it is running at negative variation.

If the shop turns out the product at less than the standard bulletin cost, it is operating at positive variation. Incorporating costs of direct labor, materials, and overhead, standard cost allows management of a particular shop to analyze their costs in each of these categories and reduce them as much as possible. Comparison of standard cost with incurred cost frequently inspires cost reduction programs that improve efficiency or reduce materials use. In summary, a company will use the standard cost to establish the price it will charge customers for a product. This approach also provides a control mechanism for analyzing problem areas, evaluating inventories, and pricing the product.

Development of Cost Measures

The development of a bulletin cost measurement system takes a considerable amount of time and expense support. In most companies changes may be introduced on a semiannual or annual basis. Like any view into the future the bulletin is designed and based on a stated set of expected conditions. Some important considerations are changing wage patterns, predicted changes in material costs, energy costs, and a myriad of other factors.

Each company will have a unique set of conditions that must be incorporated. Every item that is used in a product must be accounted for in the development of the bulletin data. Scrap or junk that is experienced at each stage of the operation must be included to insure that all costs are included. Associated costs of simple assemblies flow into the final data of large complex assemblies. The first time through, this development will be a challenge. Revisions can be made as conditions change on the existing product line. A number of computer programs that have proven helpful are available for use in this area.

RESULTS REVIEW

In most shop operations the results are analyzed on a monthly basis. At this point in time the key elements are examined and compared to target goals that have been established. If the results do not meet the planned goals, look for cost reduction in the following areas.

Analyze Labor Requirements

Compare the current man load with the forecast. Is the loading in accordance with the current schedule? If so, look one step further. Are the proper labor grades being utilized? This may be a factor in the problem. In a large shop, a few personnel working out of grade can have an impact on labor variation. In most cases the union will be the first to tell you about an employee working in a higher graded position without the proper authorization. However, sometimes employees are worked below their rated level. There are valid reasons for this. The end result will still reflect higher dollar costs in the time period under review.

Also, take a good look at machine downtime in the period. In a process operation an increase in downtime can present a distorted production picture. When this occurs the buildup of in-process hours increases and investment will follow the same upward pattern. In most systems accounting can make adjustments for labor and material.

Evaluated Investment Levels. The investment challenge to be dealt with in the shop or production environment is composed of two main components. Labor and materials are the factors that provide daily stimulation to management and engineering. The terms used to discuss investment levels may be days' stock, number of turnovers, back orders, or stock outs. The goals and objectives associated with maintaining the desired investment levels will vary with each company.

The ultimate goal would be to maintain "zero investment level." Since this is not possible it is necessary to establish a realistic level that can be anticipated and measured. Improvements in the control of inventory can be translated into cost reduction that can be readily measured. The following example illustrates one method of looking at inventory on a monthly basis.

Step 1—Production value for a 22-day work month is $440,000. This is stated at cost.

Step 2—Divide $440,000 by 22 work days to determine a $20,000 production rate per day. This is for the current period only.

Step 3—Equate material in-process and other material inventory to days of production. For example, if inventory is valued at $235,000 ÷ $20,000 per production day, the days' stock in terms of production is 11.75 days. Is this high or low?
Keep in mind that this is a mixture of labor and material since in-process (partial complete units) and raw materials are included. Elevated investment levels such as quality, shop efficiency, shipping performance, and material shortages must be dealt with daily.

Establish Your Guidelines. Concentrate on the high-volume and high-value items. In a cost reduction program a profile of these two groups is a must. Review past cost data on key items—the large runners today that are forecast to be large runners next year, also.

Determine, if you can, the cost of quality control measures associated with the cost of each product. Is it in the range of 6–9 percent? This is just a question. No implication should be drawn. However, if you can develop an answer in a short period, you may be ahead of most companies in today's marketplace.

Analyze the Material Usage

A quick way to do this would be to select the high-volume items in the production schedule. In an operation dealing with a new product, look for high-value piece parts. Ask yourself, does the 80/20 rule apply here? In most cases 20 percent of the piece parts may account for 80 percent of the dollar cost. Once the high-value items have been determined, actual purchases can be verified. This may be time consuming but could be worthwhile in resolving the current problem.

It may be worthwhile to zero in on scrap in the current period. Has something in the process changed? Has the current month served as a clearing house for material that was held in-process the month before? A minor item at first glance; however, it could be a major factor on bottom line. Most production operations have a process capability level that can be determined from quality control data. Without this information the scrap rate that is currently being experienced may not have much meaning. Are you familiar with the process capability of products in the shop that you support?

If the answer is no, this may be the time to become acquainted.

Analyze Material Buildup

The floor stock of in-process material can be equated to days' stock. Compare the value of production dollars that were generated in the current month. Divide this amount by the number of working days to determine dollars of production per day. Using this number of dollars per day, divide this amount into the inventory value on hand at the end of the period.

The quality cost under consideration should include the following:

- Prevention cost—do it right the first time.
- Ongoing cost—in-process checking, inspection, and quality assurance.

- Correction cost—detailing and rework before the product is shipped.
- Scrap material—material in a state that cannot be corrected or salvaged.
- Correction of field quality—this includes items that slipped through the "quality now" net.

COMMON MANUFACTURING PROBLEMS

Production management and engineers are appraised and usually rewarded for their achievement on a yearly basis. Some ongoing problems seem to persist that keep these managers and engineers from achieving outstanding reviews.

Four common problem areas that plague many production operations can be stated as follows:

1. Excessive in-process inventory.
2. Work flow bottlenecks.
3. Ineffective quality control.
4. Poor decision making.

Do any of these sound familiar? If you are not aware of these and the role they play in the daily routine, you may be missing out on the reality of the real world associated with the production process.

Excessive In-process Inventory

In today's environment the cost of excessive in-process inventory is a luxury that few companies can afford.

Cost Reduction Potential. Evaluate the annual savings that are possible if in-process dollars are reduced by 10 percent, 15 percent, or even 25 percent.

Select a target goal and develop your approach. Ask yourself what kind of support is needed. Set a target date to accomplish the reduction. Outline action steps that are required in order to achieve the desired end result.

Pursue your schedule and document the improvement that has been made, the dollars saved. Secure management's approval on the new method.

You are now on the road to successful cost reduction.

Work Flow Bottlenecks

Analyze the flow of material from the start to the last operation as the product leaves the shop floor. This can be done with a simple flowchart. The concern is to establish how many times a product is handled, staged, inspected, stored, and so on.

Cost Reduction Potential. Look for ways to reduce the cycle-through flow. For example, how can the present system with a cycle flow of six days be completed in four? In some cases this may not be easily accomplished with the present capital equipment and sequence of operations.

Keep this in mind: Just because something has always been done a certain way, do not be intimidated into thinking it cannot be changed.

Picture this: a shop that repairs an estimated 40 different codes of telephone sets in 10 different colors. Add to this a fleet of approximately 1500 installation trucks operated in a large distribution network. Each truck has a specified number of installations to make on any given day. The trucks were loaded the night before with certain codes as specified by the customers. No problems up to this point. A typical installer makes six to eight calls daily.

On the second installation of the day, the customer has a change of mind and wants a blue set instead of a green one. The existing service mode would allow no changes. The installer has two choices—return to the garage and pick up the desired set or reschedule the installation for another day.

No big problem you may say, a company should strive to keep the customer happy. An isolated case of this type would be no problem. In actual practice the above scenario was becoming commonplace in the mid 1960s.

The plot thickens. In order to cope, the trucks began to carry additional telephone sets in various colors. Investment increased at the same time and shop capacity was being taxed. The situation, while it was a problem, was tolerated. In early 1967 the situation was recognized as a real problem that must be resolved.

A Step Forward.   Out of the challenge came a new concept known as the Kit program. This program was built around a concept of stocking certain component parts of telephones on the installation truck. Based on historical data, a standard truck loading pattern was developed. The standard load was X handsets in colors 1–10. A contrasting number of telephone set bases in specified codes was included. The final step was that of packaging the housing, fingerwheel, line cord, and other required material in a separate package. Sounds good up to this point, doesn't it? In actual operation it was better!

The new method allowed the telephone installer to assemble a set on the customer's premises, in most cases on the first trip. In summary, the new method had all the elements of a successful cost reduction. It was that and more. This change of procedure provided an increased service level and reduced inventory, and shop scheduling was improved due to better forecasting.

Savings were substantial in terms of dollars and also customer satisfaction. The same operational mode is still in use today with further improvements. The major improvements are the ease of assembly with the new modular or easy-to-assemble plug-in components. Today's telephone installer has only to connect the handset, base assembly, handset cord, and line cord.

Customer Feedback.   As stated in the earlier chapters, one of the advantages to a cost reduction program is customer benefits. Usually translated into terms of price, quality, and service, it may not always be easy to measure.

What happened after this? The concept spread nationwide.

Ineffective Quality Control

How can you tell when your quality control program is ineffective? What are some of the indications that can confirm your suspicions? Several positive indications are as follows:

- Current quality goals are not being met.
- A lack of communication between the production employees and management is commonplace.
- A production unit flows through the line without the commitment from each team member.

- A lack of communication on quality goals or expectations may be the real problem.

The Quality Quest. The quest for a quality product has been a challenge confronting every segment of the economy. No company large or small is immune from this challenge. Many times quality goals are not met, and this affects us all. Each individual must take pride in his or her work or there will be problems with the final product. Quality requires a commitment from everyone.

When compared to most of the other 31 Western Electric Service Centers, the Southern California Service Center should produce quality products with relative ease. The service center, located near Los Angeles, is a high-volume production center. Other repair centers may run two or three different types of product during a typical workday. Even large production volumes do not guarantee a quality end product. No one individual or group can be held accountable for poor quality. The problem is everyone's responsibility.

Quality Versus Detailing. When a quality problem beyond acceptable limits is identified in a production section, a detailing form is generated. The section is now in a "detailing mode." A detailer will review the work of the section before it reaches the local inspection group. The detailer remains in a section until a specified number of units have passed all requirements without any defects.

Keep in mind that before the detailer enters the picture the repair person and process checker have made their contribution to quality. The question to be asked is, did both have an understanding of what was expected of them? Perhaps management should have conveyed what was expected in more positive terms. A daily record of errors is maintained. The person responsible for the defect can then be made aware of the problem.

Detailing is an expensive operation. Annual expense for this at the Southern California Service Center is approximately $500,000. This slowdown has an impact on efficiency and standard cost per unit.

Quality Cleanup. A quality cleanup team is a must to deal with ongoing problems. Members of this team are from Engineering, Methods, and Shop supervision. They strive to identify problems. By operating in this manner, they review all elements of the production process. The solutions may involve changes in methods, test equipment, and personnel.

This is a positive step in dealing with quality problems. However, one key ingredient is missing. The missing ingredient is the need to include production workers as a part of the team. No one is more aware of the total quality impact than an employee on the production line.

Poor Decision Making

Poor decision making by managers and engineers comes in a variety of groupings and will vary with the company. Some typical examples are listed, but by no means should this list be viewed as complete.

- Lack of planning: too busy with today's problems to look at the challenge of next week, month, or next year.
- Should we make or buy the part? Start talking like this to a production manager and check the response. Most production managers want to make everything in their own shop—even if it cannot be done.
- Buying two more of those machines at $12,000 each in order to reduce the bottleneck. The shop already had eight machines. A two-week work sampling study showed the machines were utilized 73 percent of the time. Buy more?
- Cancellation of this week's quality circles meeting due to the rush job that came in on Monday. No one will notice.
- Lack of communication with subordinates: Some managers may find it difficult to deal in a straightforward manner. This may not be popular.
- Lack of valid appraisals: Most employees expect that management will provide timely feedback and a valid yearly appraisal at a specified time. Some managers find this hard to do because they do not keep good records.
- Look at the decisions that were made last week. Would you make the same ones again?

Look to the Future. Sometimes we are so consumed with the daily problems that the long-range view is overlooked. Ask others to share ideas with you. It might be an eye-opener to see the quality of ideas that are available just for the asking.

FEBRUARY 6, 1982

RE: LONG-RANGE PLANNING

THIS IS YOUR OPPORTUNITY TO ASSIST IN PLANNING THE LONG RANGE SPACE ALLOCATION AT COMPLEX PRODUCTS, INC. PLEASE FEEL FREE TO COMMENT ON ANY PHASE OF YOUR OPERATION AS IT NOW EXISTS, ALONG WITH ANY PROPOSED IDEAS WHICH YOU FEEL SHOULD BE INCORPORATED IN LONG-RANGE PLANNING.

THIS NEW LAYOUT WILL REQUIRE YOUR ATTENTION TO THE DETAILS IN FLOW OF MATERIALS, FACIL-ITIES, ASSOCIATED STORAGE SPACE ON THE FLOOR, RECEIVALS OF SHOP MATERIALS FROM THE STORE-ROOM AND THEIR RELATIONSHIP TO OUR FUTURE GROWTH.

PLEASE LIST YOUR IDEAS IN ORDER OF PRIORITY ON THE ATTACHED PAGE.

CHIEF ENGINEER

ATT.

COPY TO:
  – MATERIAL MANAGER
  – ENGINEER
  – PRODUCTION
  – WAREHOUSE
  – QUALITY

Figure 9–7.   Long-range planning letter.

RE: LONG RANGE PLANNING

1. *MATERIAL FLOW COMMENTS*

2. *FACILITIES COMMENTS*

3. *SHOP FLOOR STORAGE*

4. *FUTURE GROWTH IDEAS*

Figure 9–8.   Long-range planning comments.

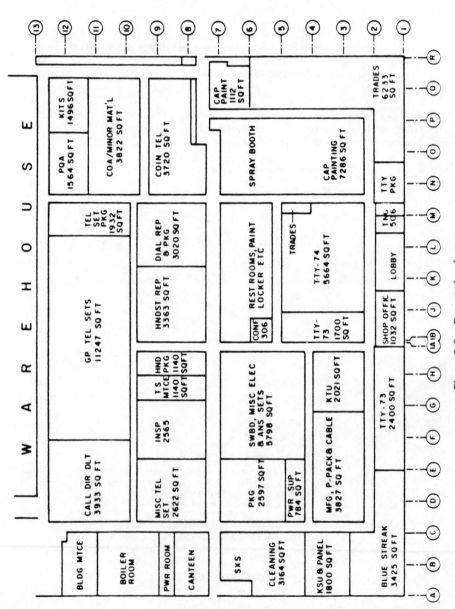

Figure 9–9. Present shop layout.

The long-range planning letter shown in Figure 9-7 may produce some valid inputs. Reply comments can be listed on Figure 9-8. The copy of the present layout can also be attached. It can serve as a memory-jogger, Figure 9-9.

Do not overlook the employee on the production line in the quest for input ideas. You might be surprised at the outcome. Look for cost reduction opportunities in the plan. Prudent planning and equating the changes to solid cost reduction will aid in selling the new plan.

In order to overcome the poor-decision-making syndrome we must be always searching for new ideas and opportunities. The following article by Ruddell Reed, Jr., offers some unique approaches to identifying productivity opportunities.

> *IE's are expected to be the productivity people.*
> *Applying the philosophy of the cost facts approach*
> *they can aim their efforts for greatest effectiveness.*
>
> Ruddell Reed, Jr.
> Purdue University

IDENTIFYING PRODUCTIVITY OPPORTUNITIES*

We have a productivity problem, have had a productivity problem, and without radical change, we are going to have a continuing productivity problem. We talk of the national problem or the regional problem, and these are important. However, most of us are going to help the national and regional productivity problems by "staying home and doing our knitting." In spite of national programs which may be developed, the real accomplishment is probably going to be made through the cumulative efforts of individual plants and organizations.

It is not simply a cliche that everyone concerned with individual plant or organization productivity is going to have to become involved. However, leadership must be provided by management and someone or some unit in the organization must be given the responsibility to identify opportunities, coordinate organized efforts, and accept the primary responsibility for program continuity and results.

---

*Reprinted with permission from *Productivity: A Series from Industrial Engineering.* Copyright © American Institute of Industrial Engineers, Inc., 25 Technology Park/Atlanta, Norcross, Georgia 30092.

IE's as Productivity People

Industrial engineers, through their professional society, AIIE, have accepted this responsibility by designating themselves, "the Productivity People." However, I think the question can reasonably be asked, "Are we willing and prepared to assume the mantle?"

I submit that perhaps during the heyday of "efficiency experts" we wore the mantle, and although we objected to the title given us by the public, it may have been our finest hour as "Productivity People." But time and conditions have changed. As efficiency experts we made great contributions in the area of direct labor.

Today, the only national measure of productivity, that of the Bureau of Labor Statistics, is output per man-hour. But direct labor accounts for only seven to twelve percent of the sales dollar in manufacturing industry according to Robert W. Semmler of Cutler-Hammer (*Industry Week,* January 5, 1976, p. 26). A limited survey by the author indicates the average may be even less, only four to eleven percent. In fact, in only one firm contacted was I able to find a direct labor cost equal to twenty percent of the sales dollar. This tells us that the opportunity for the "efficiency experts" to be the Productivity People by concentrating on reducing the cost of direct labor is past. This does not mean that the tools and techniques of the efficiency expert cannot be applied, but rather they must be applied to new arenas such as indirect labor, services, and the other eighty to ninety percent of the cost generating factors.

Some industrial engineers have recognized the limitations of concentrating on direct labor and restricting our activities to those of the traditional "efficiency experts." But in trying to move the profession forward, many people began running around with a solution looking for a problem. Many of the solutions turned out to be mathematical or computer designs for model resolution rather than problem solutions. Those people forgot what professional engineers, economists, and managers have long known. Generalizations have to be adapted to specifics, and rarely is the theory an adequate solution to the problem at hand. As a result, the models must be adapted, relaxed, and/or objectively rather than rigorously interpreted to serve as aids in resolving the real problem.

Since neither the "traditionalists" nor the "operations researcher and systems engineer" fully appreciated the contribution of the other group to his own interests, the profession was split. It is now time that we begin to bring the profession back together and combine the best of the two worlds in order that we do become the Productivity People, not only by waving a flag, but by producing as true professionals.

The Productivity Goal

Any industrial engineer's first responsibility is to begin to identify opportunities to contribute to productivity, cost, and human factors improvement. We can attempt to do this in sophisticated, idealistic terms such as total automation and optimization. In truth, total automation and optimization are long range, ideal objectives, which even in limited instances of realization are going to be accomplished in incremental steps.

Furthermore, as we automate, economics have to be considered. The level of automation which is economical today and survives the justification and funding process is likely to offer further opportunity for economic improvement by the time the installation is finally complete. The level of economic technology is shifting. If this were not true, the profession is doomed to death because its life blood is improvement.

A similar condition is true for most of the "optimal" designs. Even though we can identify an optimum by some set of criteria and variables, it is doubtful whether or not the marginal rate of return on increments of improvement before reaching the identified optimum level can be justified. In other words, the optimum solution may not satisfy the principle of maximizing the rate of return on incremental investment.

Eighty percent of the benefits are often obtained in the first twenty percent of cost. Investment in men, materials, equipment or money generally follow this rule. Therefore, we might reach a conclusion that the responsibility of "the Productivity People" is to develop a means whereby the rate of return on productivity and cost improvement efforts will be maximized over both the short term improvements. This is nothing more nor less than an attempt to optimize the budgeting of scarce resources.

Before proceeding further, and keeping in mind that we are talking about the individual organization or plant environment, we must use some measure of productivity improvement other than output per man-hour. Use of output per man-hour is like waving a red flag at a bull because it immediately connotes working harder for equal rewards.

In selecting a measure: (1) It must be distinctly measurable (which eliminates motivation); (2) It must be applicable in areas other than direct labor and direct material and across changing product mixes (which eliminates tons produced or the number of units produced); and (3) It must be applicable to all activities of the organization, and easily interpreted by all members of the organization.

In searching for such a measure the only one that comes to mind is cost. Therefore, I suggest that we have a basis for a direct relationship between cost reduction and productivity improvement so long as we recognize that we have the constraints of maintaining employment, improving the quality of working life for all individuals, and conserving scarce resources. In effect, what I am saying is that we are forced in our productivity improvement program to provide an opportunity for the individual to do a better job, with less waste, and with an increased degree of self satisfaction without jeopardizing his continued employment. Therefore, I submit that the industrial engineer's function is to find means for incremental applications of technological improvement which will result in reduced cost to the firm or improve quality of working life and in so doing maintain a maximum rate of return on his and the firm's effort under the pressures of worldwide competition.

This places the following responsibilities on the industrial engineer:

1. He must identify problems which offer opportunities for cost reduction.
2. He must order the opportunities on the basis of maximizing the rate of return on his and others' efforts.
3. He must provide leadership in determining alternative courses of action which will result in reduced cost and increased productivity.
4. He must provide leadership in predicting the cost-benefit tradeoffs of the identified alternatives.
5. He must assist in establishing priorities and programs for selected alternatives.
6. He must maintain continuity of the program by periodically revising his set and order of reduction and productivity improvement opportunities.
7. He must periodically, and as necessary, update alternative action priorities.

Ideally, in order to accomplish these objectives or requirements, we would establish a well-defined procedure with decision points identified and decision rules established. However, if such an idealized situation existed, or were easily attainable, we would probably not be in the productivity situation we are in today. Therefore, I suggest that as industrial engineers we seek to integrate "efficiency expert" and "the system engineer" concepts and methodology to establish a reasonable, logical, and workable procedure to meet the need. The approach must be adaptable to the individual firm within the constraints of that firm's unique problems and environment, including capital and personnel limitations. This eliminates running around with a solution seeking a problem because it requires identifying the problem first and then seeking a solution within the firm.

Cost Facts Approach

I choose to call this the Cost Facts Approach to productivity improvement. The underlying philosophy is the application of the Pareto or Value Curve to avoidable costs. The logic is to first identify and estimate avoidable costs, second, establish problem priorities on the basis of the relative magnitude of the estimated avoidable costs, and third, to seek means to reduce the avoidable costs in magnitude order priorities. The result will be increased or equal output at reduced cost.

To the organization it should be immaterial whether the Bureau of Labor Statistics' measure of productivity is improved if the organization effectiveness is increased and competitive position reinforced. Also, by ordering the avoidable costs and establishing priorities by that order, we increase the likelihood of maximizing the return on the effort expended.

Two major problems arise early in this approach. First, we have no direct method for identifying avoidable costs. Accounting records, which should be our primary source of basic information, are not designed to identify causes of cost but rather categories or areas of cost. This results from accounting's primary responsibility normally being defined as providing a profit and loss statement while satisfying the constraints and regulations of the Internal Revenue Service. Also, budgets are established and controlled on the basis of management and supervisory responsibility.

This accounting record problem cannot be overcome without radical changes in basic cost input methods and data. Since this is not likely to take place, at least in the short run, substitutes or alternatives for identification of avoidable costs must be attempted. Some possible methods of identifying potential avoidable costs will be covered in a moment.

The second major problem is in identification of alternatives to reduce an identified avoidable cost. Perhaps the greatest danger here is that the analyst will consider only two alternatives; keep things as they are, or seek an idealized solution. The economical and feasible alternative approvable by management probably lies between these two extremes.

To establish a standard or formal method for identifying intermediate alternatives, use can be made of a precedence network to identify the actions or events necessary to reach the ideal solution from the present condition. It is highly likely that such a precedence network will identify intermediate levels of improvement which can be acomplished and permit incremental analysis of cost-benefits. An advantage of the precedence network is that it provides a systems approach assuring that the intermediate steps or alternatives will contribute to reaching the ideal goal with minimum deviation. This minimizes the likelihood of regressive or duplicate action.

As an example, much attention is presently being given to computer aided design and computer aided manufacturing (CAD/CAM). However, it is unlikely that the majority of American firms can afford the expense of installing and debugging "state of the art" systems within the firm's cash and cost recovery period constraints. At the same time there is reason to believe that the major recovery may not be at the machine and tool point but in the storage, handling, and delay portions of the total process cycle since actual operation time accounts for one and one-half percent or less of the total processing time.

The in-process material handling, storage, and control system, including the reporting system, may be improved to increase this one and one-half percent to five or ten percent (and conceivably even more) at far less cost and with less disruption of operations. After increasing the percentage of total in-process time during which operations are being performed, incremental cost analysis, capital budgeting, and resource allocation methodologies may be used to develop a program for conversion of operations to automation.

Identifying Costs and Priorities

Returning now to the primary problem of identifying and ordering avoidable costs, examine Table 1 and 2. It indicates areas of potential problems with some potentially applicable techniques to be used in problem resolution. This list is definitely incomplete. Neither can a general ordering of potential avoidable costs be given since the order is unique to the facility under study.

No list is complete nor applicable to all facilities. While generalizations may be made, specifics must be uniquely developed. Only the facility can identify its own areas of avoidable costs. Nor is it likely that any unit of the organization of that facility can adequately identify all potential improvement areas. Not even the IE's. Therefore, some means must be found to obtain the inputs of a range of interests and knowledge initially. It should also be kept in mind that workers, as well as professionals and managers can contribute.

Since the most difficult stage is to identify the problem and estimate the potential benefit from eliminating or reducing avoidable costs to establish their position in the Pareto or Value Curve, we must seek means of getting this input from many sources. Techniques available for this include questionnaires, personal interviews, value analysis reports, and cost reduction proposals. All these and others can help, but perhaps most important is to get the IE's to keep their ears open and tuned in for casual comments or "bitches." A little followup and inquiry may well offer an item which should

Table 1. Identifying and Resolving Problems.

PROBLEM AREA	TECHNIQUES TO HELP IDENTIFY THE PROBLEM	TECHNIQUES TO HELP RESOLVE THE PROBLEM
In-process inventory	Activity sampling	Fixed path handling Move ticket control Fixed storage location Finite scheduling Capacity determination Group technology
Information timeliness and accountability at point of use	Activity sampling Control charts	Process analysis of input and output data external to the computer Form and report design Routine operating system
In-process queues	Work sampling Control charts	In-process inventory control Timely, accurate, and usable information Control charts
Job, material, tool, machine delays due to one or the other, e.g. Lack of adequate coordination	Sampling ANOVA	MRP (materials requirements planning)
Maintenance	Sampling	Methods study Time standards Spares program Simulation
Damaged parts, scrap, and rework	Sampling Design of experiments ANOVA	Handling Storage Tooling Low cost automation
Expediting	System analysis	Information timeliness and accuracy MRP Planning, scheduling, and dispatching algorithms Material flow systems design

be added to the list of avoidable costs for either immediate or future attention. In any case, the list must be formal and continuously maintained.

In summary, if industrial engineers are to be the "Productivity People" they must take the leadership in identifying, analyzing, and justifying courses of action in their present assignment. Few will be able to secure support for "state of the art" type proposals, which are largely unproven in practice, and all will be constrained by cost vs. capital availability. Most progress will be made in relatively small, incremental steps. But collectively these incremental improvements will provide the productivity improvement

Table 2.   Identifying and Resolving Problems.

PROBLEM AREA	TECHNIQUES TO HELP IDENTIFY THE PROBLEM	TECHNIQUES TO HELP RESOLVE THE PROBLEM
Short run set-up cost	Sampling	Group technology
		N/C sequencing
Assembly delays	Sampling	Group technology
		Modularization
		Cellular manufacturing
		Methods analysis
		Lost cost automation
Distribution	Cost and time delay	Simulation
		Linear programming
		Network analysis
		Dock design
		Order gathering
Conservation of energy	Survey sampling	Storage area design
		Handling systems
		Equipment replacement studies
		Energy use audits
Stores or finished goods	Sampling	Systems and procedures
		Information systems
		Inventory aging
		Modularization
		Group technology
Material handling	Activity sampling	Centralized dispatching
		Fixed-path equipment
		Semi-automation
		Code reading systems
Clerical	Work measurement	Forms design
	Sampling	Coding
	Methods analysis	Semi-automation
		Data gathering
		Data processing
		Job cost reporting

critical to the future well being of American industry. To make this incremental improvement requires use of both traditional and more modern methods and technology. But before technology can be applied, problems must be identified and put in order of priority to maximize the benefit of individual efforts. The Cost Facts Approach to productivity improvement provides a vehicle.

10
Warehouse Operations—
Savings Unlimited

In the total production and distribution cycle, the warehouse function in most companies is probably the least cost effective. There has been a delay in the application of warehousing technology. However, management is beginning to acknowledge the fact that warehousing is an area that can provide fruitful cost reduction. In some companies this potential is now being developed. The current challenge for warehousing management is to recognize the need for a cost-effective operation and utilize the available tools to that end. In order to improve operating results in the warehousing environment, proven tools and techniques must be introduced. However, the improvement tools must be compatible with the stated goals and objectives. The bottom line is that the warehouse must become more cost effective without a sacrifice in the service area. Effective cost controls are the key element.

The objective of warehousing cost controls is to keep expenses to a minimum compatible with the overall objectives of the warehouse. Cost controls are effective only in relation to the warehouse's objectives. What warehousing management intends to achieve in these areas will largely dictate the levels of necessary expenditures. Use of effective cost controls will provide a means of keeping the expenditures near these established levels. The warehouse objectives serve as a frame of reference within which cost controls are established and maintained. Without objectives, the cost-control program has neither direction nor reference.

To appreciate the need for warehousing objectives in an effective cost control program, visualize the difference in expense levels. Assume that a retail warehouse deals mainly in small-volume orders which require repackaging and a lot of handling per unit shipped. The handling cost per unit should be higher than it is for the storage warehouse, which is used only to store and ship full pallet loads. The level of expenses for the two warehouses per unit shipped will be, and should be, much different. To expect both warehouses to have the same costs for the same volume would be unrealistic. The same differences exist in customer-service requirements. Two warehouses storing and handling the same product mix will, and should, have different expenses if their order-processing time requirements are substantially different. If one warehouse is required to provide same-day service to its customers, its costs should be higher than those of the warehouse that has two days to fill and ship an order. In general, cost control systems can be simple or complicated. In most warehousing operations, the system need not be elaborate to document results, although in some large production operations it must be very sophisticated to be at all effective. Generally, warehousing has fewer functions and cost classifications than have operations that involve assembly and fabrication.

EXPENSE FACTORS

In broad terms the warehouse function can be defined as a nonvalue-added operation. Say you cannot accept this basic definition? Give this some thought and you may change your mind. Most warehouse operations consist of the following six basic elements:

- Receive material into the warehouse. The two most common modes are truck and rail.
- Verify material. This is accomplished between the receival and put-away operation. The most common technique is referred to as 100 percent verification. Other common modes may include a sampling technique.
- Put away material. Normal routines include the movement of material by a hand truck, a fork truck, or perhaps a dragline, or other more modern carriers. Once the material has been moved, dropped, staged, sorted, it is then placed into some storage mode. The storage mode may be a pallet rack, a shelf unit, or a floor pile for bulk material that can be stacked.

So far we haven't added any value to the product, right? From this point it doesn't become any better!

- Select material. This is the next step. This step occurs when an order has been placed. In a manufacturing environment the material may be selected for movement into the shop, which could be 1200 feet away, or for a customer in the same city or miles away.
- Stage material. This step is accomplished prior to the material's being shipped. At this point the material is verified, checked against the order, packed, sealed, labeled, and secured for shipment. This phase of the operation varies widely with the product, industry, and shipping mode. At this point the material is ready to ship.
- Ship material. After the staging, the material may be loaded directly on a truck, picked up by United Parcel, or shipped Air Express. Regardless of the mode selected, the expenses continue to mount. This is true if the material is moved only 600 feet to the start of an assembly operation.

After the material has been shipped from the warehouse, the expense incurred could be stated in several different ways. The dollar cost per unit may be a valid measure in some operations. Other situations may find that a reference to cost per case or dollars per $100 of material may be useful. Regardless of how it is stated, warehousing is a very expensive operation.

Up to this point you may have noticed that no mention has been made of the quality aspects of the operation. This endeavor can prove to be a major challenge.

Warehouse Expense Factors

Develop a profile of warehouse expense factors for your operation. Keep it simple yet make sure that you include all expense inputs.

Analyze the current operation in terms of space, labor, facilities, and other expense. Let us review each in order to insure that a common starting point is established.

- Space: should include both owned and leased space. This should also cover heat, lights, water, security, and other related-to-space expenses. Be sure to include depreciation.
- Labor: should include direct labor and indirect labor.
- Facilities: should include equipment depreciation, maintenance

charges, and other related items. All types of equipment should be included—pallet racks, shelving units, lift trucks, mezzanines, and other related equipment.

• Other expense: warehouse supervisory salaries, pallets, slip sheets, packing material, labels, and other required expense items. Also, any leased equipment—fork trucks, etc.—should be included here.

After the profile has been developed you now have a picture of the expenses associated with the present system. From this point it will be possible to examine all factors for cost reduction potential. In the review it will become apparent that some areas present more opportunities than others. From this point it is possible to explore areas that will lead to methods changes in the operation that will reduce expense.

PROFILE REVIEW

The profile has been completed and the best data at this point in time is as follows in this hypothetical case.

Space	Owned and leased	30%
Labor	Direct and indirect	50%
Facilities	Depreciation, maintenance, and other	5%
Other expense	Supervision and expense materials	15%
	Total	100%

What now! It is probably no big surprise that in most warehouse environments labor is the major expense element. In this example space is also a major element. Facilities and other expense make up the balance. With the profile as a guide, let us develop a probe list for each grouping of expense. The labor component in most cases is defined as the really weak key to improving productivity.

The probing cost reduction approach may start in the following manner. List items for review in each category of expense.

1. Space improvement ideas can contribute to cost reduction.
   1.1 List all locations owned and leased. Break down each by size function, based on the allocation of space.
   1.2 Break down the space as it is currently assigned into square feet. Your list will include receiving, staging, rack storage, aisles, mez-

zanine, staging areas, office space, toilet areas, drop areas, and shipping dock. Be sure to include any other areas that are unique to your operation. Refer to Figure 10–1.

Some of these areas will remain fixed and others may shift in size to provide increased potential for meeting service needs. The overall objective is to assign the available space in such a manner that productivity can be increased and measured to document. This is our end result in all cost reduction efforts.

1.3 Review current utilization of cubage. In addition to space usage in a two-dimension plane, the maximum use of the total cube space is essential. In what may look like a prudent assignment of space the calculation of the cube in a normal warehouse may be a shocker! Cube utilization in the range of 15–20 percent may not be unusual.

2. Labor utilization is the area that presents the most unique challenge. One of the first questions to ask is, how were the present man-load requirements determined?

2.1 Predetermined time standards offer a technique that can meet the constant changes in daily operations. They provide a way of evaluating proposals that may lead to improved productivity. How could you document the change in one select proposal in favor of another without firm data?

Do you currently use predetermined time standards?

2.2 How can current labor-intensive elements of the operation be restructured through the use of mechanical means? This is a good question and may provide an entry into valuable savings. Consider the following.

Can a dragline deliver material to the shipping dock versus a forklift truck with a driver?

Would a "smart cart," a driverless pallet-moving device that tracks a wire in the floor, be a good replacement for the truck and driver combination? The "smart cart" can be directed to stop at or bypass a number of load and unload points in the fixed closed-loop configuration.

○ PALLET STORAGE AREA
○ SHELVING STORAGE
○ FLOOR STORAGE
○ AISLE SPACE
○ SELECT, STAGE AND PACK
○ SHIP AND RECEIVE AREAS
○ OFFICE AREA
○ GROWTH SPACE

Figure 10–1.   Development of facility size.

2.3 How much labor do you require in verification of material coming into the warehouse?

Shipments from outside suppliers may need to be verified 100 percent of the time. However, shipments from one location to another in the same company could be accepted with a random sampling acceptance technique. An example of this would be a random sample to validate a truckload of telephone sets from the Indianapolis factory that are received at a regional material management center.

Another level of material acceptance might be a material manager in a large central warehouse that ships daily to a number of smaller strategically placed warehouses. The same manager is responsible for profit and loss, service, and inventory levels in both cases. A reduced level of inspection could be applied here. Why incur the expense of 100 percent verification when the material has never left the organization? This is not a small item. Investigation may show that this equates to one or more people.

Expressed in dollars on a yearly basis the cost cannot be overlooked. How does your company handle material that is received in the warehouse?

List other ideas on how to get a better grasp on labor expenses. Talk to other supervisors and engineers and see if they can share an idea with potential for cost reduction development.

3. Facilities associated with the warehouse offer an opportunity for cost reduction.

   3.1 Look at the ongoing maintenance of equipment. Do you plan specified maintenance tasks as a routine matter or just when the item breaks down?

   3.2 Conduct utilization studies to determine how many trucks are really needed for the operation. Surplus trucks can usually be transferred to another location in a big company. Each owned truck that is removed will have a positive impact on the depreciation charges.

   3.3 Examine the techniques of maintaining batteries for trucks in a proper manner. Lack of attention in this area can be expensive.

4. Other expense offers a wide variety of entry in cost reduction. First take a good look at the major expenditures in the group.

   4.1 How many supervisors do you really need to conduct the daily operation? Review your reporting structure. Can you eliminate one level of management? A number of experiments have been tried in this area. Some have been very successful.

   4.2 Take a good look at packing material. Do you feel that too many different carton sizes are stocked? Review usage patterns for the last six months. How about dead stock? The codes can be removed. Check for dust on infrequently used codes. With a little effort it may be possible to achieve a 25 percent reduction in the number of carton codes that are stocked.

4.3 How about labels versus an automatic carton-marking device? Outside supply sources for labels may offer an economy if labels are needed. However, if a large variety of different labels in small quantities are needed, it may be more economical to generate them in-house.

4.4 Pallet control techniques offer a fertile field for cost reduction. Ask yourself, how many trips does the average pallet make? Should you consider using a disposable pallet? How about billing the customer for the pallet and letting disposal be the customer's consideration? How about renting pallets? Not as far out as it may seem. How would like to rent a pallet for $2 from point A to point B? It is now possible to do this under a new offering. More data on this item are given later in the chapter.

A DISTRIBUTION STRATEGY REVIEW

The number of distribution warehouses in the United States has shown substantial growth in the period 1965–1980. This has certainly been the case in Western Electric Company, along with the others that operate on a national basis. A recent study by Handling and Shipping Management indicates the following trend changes.

In 1965 the number of distribution locations per U.S. company was pegged at 3.56 locations. By 1975 the typical company had 9.68 distribution locations. Again in 1980 the trend was upward to 10.68 locations. The pattern of growth has been impressive, but will it continue? The trend was the result of thinking, "Service for the customer." Will the present economic uncertainties of high interest rates, increased energy costs, and transportation deregulation slow the expansion? The answer can be a firm "yes" at this point.[1]

COST REDUCTION POTENTIAL

The current challenge for the long-range planners is to evaluate the economic trade-offs in the market place today. Let us consider a few that may be worthwhile.

•Own or lease the warehouse building? Each case is different. Past history indicates that many large companies have chosen the ownership mode. This may not be true today.

A lease, on the other hand, offers a flexible plan to keep pace with business needs. Population shifts, distribution costs, and the product itself may dictate that a five-year lease with five one-year options is a better involvement.

Look at the major good chain stores, Safeway, Alpha Beta, Big Star, Colonial Stores; leasing is the most common. If ownership were the mode of operation, these companies could soon find that they were in the real estate business versus food distribution and sales.

•Buy or lease pallet racks and shelving units? This is a new concept that has a lot of potential in these days of tight money. Most companies, large and small, are in the same dilemma. Recently I worked on a proposal of this nature to equip a leased warehouse that was in the planning stages.

Each supplier that quoted on the job was instructed to furnish a price for a standard purchase agreement. Also, an option for a 60-month lease with a specified buy-out at the end of the period was requested.

To illustrate this example, a standard agreement to purchase the equipment was approximately $211,000, payable in 30 days. A lease for 60 months could be obtained at a cost of 18 percent APR, or $25.39 per $1000, for a monthly payment of approximately $5357. Any advantage to this concept, you may ask? The main consideration is that many companies may be able to cope with the lease cost versus the standard approaches of direct purchase. One main advantage is that the monthly payment for the lease can be assumed as a current operating expense. Also, a buy-out for a small sum at the end lets the company retain the racks, through a minimum purchase agreement. One word of caution: Obtain a current legal opinion before entering into any agreement of this nature.

•Buy or lease warehouse trucks and pallets? This problem offers some interesting trade-offs.

Past practices may have shown that a mixture of owned versus leased trucks was beneficial. Leased trucks could have been obtained for peak periods. These peak periods vary by industry type and the stage of company development. A peak period for a large canning company in central California may be July through September. A large merchandise company may have a November through January peak period. Look at your company. Can you identify any peak periods?

The decision to lease or buy may be influenced by the time period under consideration. In other words, a one-year versus a five-year lease could cost twice as much per month. This was emphasized as I discussed both time periods with a truck supplier in the St. Louis area. For example, a reach truck to operate in a 7½-foot aisle could be purchased for $23,850. The lease charge for a 60-month period would be $697 monthly, with a buy-out of $3800 at the end of the period. On the other hand, a one-year lease would be $1394 per month. Which offer is best? That question has no one answer. The time frame of the answer is the key. As stated before, before entering any such agreement secure a legal opinion. Some leases may also provide a tax benefit. Check these claims very carefully.

Let us address the common warehouse pallet at this point in the review of warehouse expenses. At first glance it may appear that this could be a short discussion since pallets do not account for many expense dollars. If you were prone to reply in that manner, very possibly you do not really have a feel for the real dollar cost linked to pallet control.

For example, let us say that a company has a well-integrated manufacturing network in the Midwest and East, but a large percent of sales are on the West Coast. To meet this need a large warehouse in the range of 600,000 square feet is located in the Sacramento, California, area. Products from plant locations are shipped via truck and rail into this distribution center. Up to 30,000 different items could be stocked in this distribution facility. The material is received and stored on pallets. Approximately 35,000 pallet positions are available for storage of material. High-rise shelving and bulk floor storage are also utilized.

Daily shipments are made to Seattle, Portland, Oakland, and Los Angeles. At these locations the problem of pallet accumulation begins to develop.

How to recycle the pallets back to the original location is the challenge that must be met. No problem, you may say, just load the empty pallets, 350 or so per truckload, and ship them back. Sounds simple, and it is. The problem comes into focus when the freight bills start rolling in a few weeks later. Keep in mind the freight rates quoted are subject to a fuel tax surcharge of approximately 18.5 percent.

In round numbers it may cost approximately $8 each to return a

STW NATIONAL PALLET LEASING SYSTEM

INTRODUCES

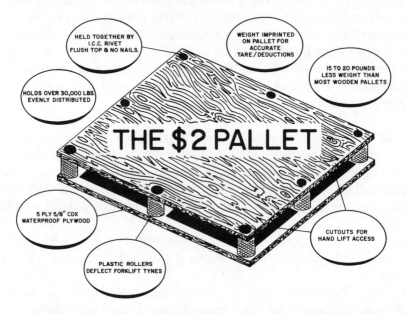

HELD TOGETHER BY
I.C.C. RIVET
FLUSH TOP & NO NAILS.

WEIGHT IMPRINTED
ON PALLET FOR
ACCURATE
TARE/DEDUCTIONS

15 TO 20 POUNDS
LESS WEIGHT THAN
MOST WOODEN PALLETS

HOLDS OVER 30,000 LBS
EVENLY DISTRIBUTED

THE $2 PALLET

5 PLY 5/8" CDX
WATERPROOF PLYWOOD

CUTOUTS FOR
HAND LIFT ACCESS

PLASTIC ROLLERS
DEFLECT FORKLIFT TYNES

WHY BUY A PALLET WHEN YOU CAN LEASE ONE FOR $2!

Figure 10-2.   Rent a pallet.

truckload of pallets from Seattle to Shreveport. Similar charges of approximately $4 can move the pallet from Los Angeles to Phoenix. When compared to an original purchase price around $10 each for the pallet, return charges alone can run from 40 to 80 percent of the first cost.

A recent innovation is that of leasing a pallet for $2 plus a deposit. The deposit can be obtained after the pallet has been used to ship from point A to point B. Such a service offering is now available from SYSTEMS THAT WORK in Fairfax, California. Refer to Figure 10-2. When you have finished using the pallet, it is picked up from your warehouse in truck lots, say about 350. The deposit less any damage charges is returned.

This item may have cost reduction application in your situation. Prepare an economic study; if the results look good, your next step is selling the idea to management!

WHERE WE ARE NOW

During the mid-1960s many long-range planners and seminar speakers had said that a high percentage of the national consumer markets could be serviced within 48 hours by having three to five well-placed distribution centers. Also, they maintained that it would be possible to create favorable economic trade-offs between transportation costs and inventory carrying charges. At that time they were correct.

Inexpensive transportation was traded for field investment. This was due to available transportation and cheap energy. The desire to protect customer service levels led to further expansion of the distribution network in the 1970s.

Enough history at this point. Where do we go from here?

What are the key factors that will shape our future? Three factors can be identified as having a major impact.

1. High interest rates, which may be around for a long while.
2. Escalating fuel costs coupled with the concern of fuel availability.
3. Concern about the cost and availability of adequate transportation in an unregulated environment.

With this in mind we must learn to be more productive with the resources that are available. In this new environment material inventories will require better management than before. Labor must be utilized to provide expanded service by working in a more productive manner. Learning to control inventory and labor will be our major challenge in the 1980s.

REVIEW OF WAREHOUSE OPERATIONS

How can we compare one warehouse operation against another? Which partial productivity measurements can best illustrate a productivity trend? This question could be asked of several experts in the field. The feedback might be similar and yet slightly different.

To set the stage for our evaluation we will select a company that has a well-developed network of distribution. Our hypothetical company will be Complex Products, located in Raleigh, North Carolina. This is the same company that was illustrated in Chapter 9. The objective of the study is to compare the operating modes of the small distribution centers. A total of 12 locations are operating in the state of California.

Management control is split between northern and southern California. Locations of the small distribution warehouses were determined by In-House Industrial Engineering studies in conjunction with an outside marketing consultant. The size of each warehouse ranges from 25,000 to 40,000 square feet. The key in planning the network was directed at improving service to the customer.

The last of the 12 warehouses was opened in October 1979. Sales, service management, and inventory all appear to be tracking the forecast budget that was established for 1980. In summary, when viewed by upper management the operation appears to be performing in an acceptable manner.

The Review and Comments

At the recommendation of the industrial engineering manager a midyear review was planned. The objective of the review was to determine the most efficient and least efficient operation. The review was prepared by a team of two skilled engineers after a visit at each warehouse location. A letter was prepared to inform management of the results.

Refer to Figures 10-3, 10-4, 10-5, and 10-6 for the results of the engineering review. This data clearly indicates the need for more study and evaluation in order to determine why the measurements vary to such a wide degree.

An action step is stated in the letter from the industrial engineering manager, Figure 10-3. Support from management at this point is essential in order to proceed into the study phase. This should be no big problem with the data shown in Figures 10-4, 10-5, and 10-6. A prudent management team would be challenged by this review of productivity measurements. Some valid questions can be asked at this point.

- How can the selected measures vary this much in the same basic operation?

September 10, 1980

Material Distribution Manager
Re: Warehouse Data Profile and Measurements
Attached for your review are three data sheets which provide a comparison of the twelve operations for the month of July 1980.

Several problem areas are apparent from a review of the data.

a. Hours of operation vary widely from one warehouse to another.

b. Confirming orders are, on the average, significantly higher in the Northern Region and require additional clerical and warehouse labor to process the paperwork. We understand that confirming orders also cause computer balances and back order routines to become distorted.

c. Labor costs per line item or $100 of material shipped vary considerably.

d. Operational modes vary from one warehouse to another. An example is the delivery load truck used in the Northern Region; also, loading and unloading at some locations differs widely.

In summary, the variance in service measurements can be attributed to different modes of operation. From the attached chart (Productivity Measures) San Diego is rated the most cost effective and San Francisco the least, based on the six measurements shown. The maximum score possible is 72. The minimum score that can be achieved is 6. Each of six items selected for evaluation were assigned a point rating of 12 for the most effective and 1 for the least effective in all cases except for average dollar value of a line item. In this case, a reverse ranking was used by assigning a 12 to the lowest value and a 1 to the highest value. Points were totaled for all six measures to determine a composite total for each. These are summarized below:

LOCATION	POINTS	LOCATION	POINTS
San Diego	62	Fresno	39
Monterey Park	56	Sacramento	28
Van Nuys	56	Modesto	24
Tustin	54	East Bay	23
Culver City	53	Vallejo	21
Santa Clara	41	San Francisco	11

The wide disparity of operations highlights a need to work toward standardization of operations. It is our recommendation that labor standards be developed and introduced at each location.

Our current evaluation indicates that four industrial engineers can develop the required labor measures in a six-month period. I would like your support in this matter at our next weekly staff meeting.

Industrial Engineering Manager

Figure 10-3.   Warehouse data profile and measurements letter.

MEASURE	NORTHERN	SOUTHERN	VARIANCE
1. $ shipped per hour worked	$976	$640	52%
2. Line items shipped per hour worked	10.6	6.5	63%
3. Line items shipped per employee	1476	907	63%
4. $ shipped per employee	$135,576	$89,560	51%
5. $ labor cost per line item	$1.71	$2.72	59%
6. $ labor cost per $100 shipped	$1.85	$2.76	49%
7. Avg. $ value per line item	$91.86	$98.73	7%
8. Avg. labor rate	$18.12	$17.66	3%
9. Confirming orders	53%	81%	59%
10. Direct employees	103	174	69%
11. $ shipped total ($000)	$13,964	$15,583	12%
12. Total line items	152,009	157,833	4%

Figure 10–4.   Warehouse profile, July 1980.

- How can these differences be translated into cost reduction savings?
- What are the potential savings that can be expected from the study?

Organizing the Study

The study as designed by the Industrial Engineering group was stated at the weekly staff meeting in the following manner:

- Each required work function will be covered in a job description.
- Descriptions of related work activity will be covered in detail. Each activity can be analyzed by means of a flowchart that contains each element and the distance traveled in order to complete the activity.
- Methods to accomplish the work in the most efficient manner will be documented. The constraints will be the building, the layout, and material handling equipment.
- Frequencies for each activity will be determined.
- Labor standards will be calculated on the basis of the job description, the scope of work, the most efficient method, and the related frequencies.
- The labor standards multiplied by average daily volumes will provide the standard hours that are required to accomplish the job.

The end result will produce labor standards that can be converted to the required man load for each warehouse. The reduction in man-load requirements will be translated into cost reduction savings.

SOUTHERN REGION	$ SHIPPED/HR	LINE ITEMS/HR	$ COST/L.I.	$ COST/$100	$ AVG./L.I.	$ SHIP/EMP.	POINTS
Van Nuys	766 (7)	10.9 (11)	1.71 (11)	2.44 (7)	70.05 (11)	111,156 (9)	56
Culver City	950 (10)	9.6 (8)	1.91 (10)	1.93 (10)	98.99 (5)	135,684 (10)	53
Monterey Park	1101 (11)	10.2 (10)	1.84 (9)	1.71 (11)	107.59 (4)	152,280 (11)	56
Tustin	1249 (12)	9.9 (9)	1.72 (8)	1.36 (12)	126.30 (1)	168,937 (12)	54
San Diego	814 (9)	12.2 (12)	1.44 (12)	2.16 (9)	66.81 (12)	110,353 (8)	62

NORTHERN REGION	$ SHIPPED/HR	LINE ITEMS/HR	$ COST/L.I.	$ COST/$100	$ AVG./L.I.	$ SHIP/EMP.	POINTS
Santa Clara	767 (8)	7.8 (6)	2.18 (7)	2.21 (8)	98.74 (6)	101,883 (6)	41
San Francisco	492 (1)	4.3 (1)	3.94 (1)	3.41 (2)	115.71 (2)	83,240 (4)	11
East Bay	594 (4)	6.2 (2)	2.84 (3)	2.94 (4)	96.38 (7)	82,195 (3)	23
Fresno	718 (5)	8.0 (7)	2.39 (6)	2.67 (5)	89.33 (9)	102,738 (7)	39
Modesto	564 (2)	7.5 (5)	2.57 (5)	3.43 (1)	74.87 (10)	72,856 (1)	24
Sacramento	724 (6)	6.5 (4)	2.80 (4)	2.53 (6)	110.82 (3)	93,267 (5)	28
Vallejo	593 (3)	6.4 (3)	2.90 (2)	3.17 (3)	91.53 (8)	79,082 (2)	21

Figure 10-5.   Productivity measures. (Facility costs are not included in the data.)

LOCATION	EMPLOYEES	HOURS	$ SHIPPED	LINE ITEMS	CONFIRMING	HOURS OPEN	HOURS DAILY
Van Nuys	18.0	2,609	2,000,813	28,560	83	7:00 A.M.–12:30 A.M.	17.5
Culver City	20.0	2,856	2,713,678	27,412	23	4:30 A.M.– 8:30 P.M.	16.0
Monterey Park	16.0	2,212	2,436,478	22,645	30	4:30 A.M.– 8:00 P.M.	15.5
Tustin	24.0	3,247	4,054,487	32,101	65	5:30 A.M.– 9:30 P.M.	16.0
San Diego	25.0	3,388	2,758,837	41,291	57	5:30 A.M.–11:00 P.M.	17.5
Total	103.0	14,312	$13,964,293	152,009	53% Avg.		
Santa Clara	38.0	5,047	3,871,583	39,208	78	5:00 A.M.– 1:00 A.M.	20.0
San Francisco	29.0	4,898	2,413,963	20,862	83	6:00 A.M.–12:30 A.M.	19.5
East Bay	32.0	4,426	2,630,247	27,291	86	6:00 A.M.–Midnight	18.0
Fresno	17.0	2,432	1,746,543	19,550	86	6:00 A.M.–Midnight	18.0
Modesto	13.5	1,744	983,567	13,136	88	5:00 A.M.–11:30 P.M.	18.5
Sacramento	29.5	3,799	2,751,367	24,827	78	7:00 A.M.–11:30 P.M.	16.5
Vallejo	15.0	2,001	1,186,243	12,959	65	6:30 A.M.–11:30 P.M.	17.0
Total	174.0	24,347	$15,583,513	157,833	81% Avg.		

Figure 10–6.   Warehouse data sheet.

The study can be completed by four engineers in a six-month period. Estimated cost for the study includes salaries, travel and living, and other related expenses. Total development expense for the study is estimated at $125,000. Sounds like a lot, you may say? This is only one side of the story.

Return on Investment

Like any endeavor, the return on investment is a consideration that must be evaluated. At the start of the study a total of 277 personnel were assigned to the warehouse network. This number reflects all employees. A reduction of 10 percent is established as an initial goal. This has been shown as a reasonable goal based on past studies.

A reduction of 28 employees will produce a sizable expense reduction. The dollar projection is estimated to be in the range of $915,000. The rate of return by any method of calculation will be substantial. The real advantage to this type of cost reduction case is that no large outlay for capital equipment is required.

The Completed Study—Six Months Later

When completed, the study will provide management at each warehouse with a technique to develop labor requirements.

Refer to Figures 10-7 and 10-8 for the summary sheets that generated from the study. The key items on each summary sheet are the monthly volumes, the equivalent people on roll, and the industrial engineering man load recommendation.

This data can be used to adjust the man load. One key point: Do not expect the reductions to be made the instant the study is completed. It may be easier to sell the reductions in one, two, or even three phases in order not to disrupt the operation. Once the standards are available, the supervisor at each location should define what is expected from each employee. This is done in conjunction with an explanation of the associated methods changes.

Close coordination between engineering and management is required in the decision-making process to adjust from the present workforce down to the recommended work force. An engineer involved in this type of program should have the facts to aid in selling his idea. Selling the idea leads to the action step. The action step, in this

SDC	SEPT.	OCT.	NOV.	DEC.	JAN.	FEB.	MAR.
Sacramento:							
Volume	2659	2661	2465	2286	2505	2547	2671
Equivalent People	30.5	26.0	24.5	24.5	27.0	26.0	25.0
I.E. Manloading	19.5	19.5	18.5	17.5	19.0	19.5	19.5
San Francisco:							
Volume	1757	1586	1566	1521	1591	1663	1780
Equivalent People	30.5	28.0	27.0	25.0	23.5	24.0	23.5
I.E. Manloading	16.0	14.5	14.0	13.5	14.5	15.0	16.0
Vallejo:							
Volume	1422	1464	1354	1263	1257	1346	1348
Equivalent People	15.0	14.0	14.0	13.5	13.0	14.0	14.0
I.E. Manloading	12.5	13.0	12.0	11.0	11.0	12.0	12.0
Oakland:							
Volume	2246	2419	2397	2198	2567	2673	2645
Equivalent People	33.0	30.5	28.0	31.5	28.0	28.5	27.5
I.E. Manloading	19.5	21.5	21.0	19.0	22.5	23.5	23.0
Modesto:							
Volume	1165	1250	1189	1061	1208	1183	1173
Equivalent People	12.0	12.0	18.5	14.0	12.0	12.5	12.0
I.E. Manloading	13.0	13.5	13.0	11.5	13.5	13.0	13.0
Santa Clara:							
Volume	3191	3286	3078	2907	3211	3267	3339
Equivalent People	36.5	31.5	32.5	32.0	30.5	30.0	32.0
I.E. Manloading	27.5	28.0	26.5	25.0	27.5	28.0	28.5
Fresno:							
Volume	1827	1825	1851	1863	1939	2023	1981
Equivalent People	17.5	17.0	15.5	14.0	15.5	14.5	14.0
I.E. Manloading	11.5	11.5	11.5	11.5	12.0	12.5	12.5

Note: Encircled data represents month during which study was conducted.

Figure 10-7. Industrial engineering study, northern region. (Encircled data represent month during which study was conducted.)

case, is the man-load adjustment downward. When this is accomplished the cost reduction case can be closed with savings. The data can then be submitted to management for final approvals.

The use of a graphical presentation, as shown in Figure 10-9, can be an asset in selling the idea to a successful conclusion. The graph illustrates the relationship between line items on a daily basis compared to manpower to handle that level of business.

The increased level of productivity is evident. An equivalent 29.5 people were required to handle the level of business in July. A reduction of 5.5 people, down to 24 people, is now processing more line items on a daily basis than before. This equates to a true increase in productivity.

How many dollars is this reduction worth? In a warehouse environment the 5.5 people, based on direct labor and fringe benefits, may

SDC	SEPT.	OCT.	NOV.	DEC.	JAN.	FEB.	MAR.
Tustin:							
Volume	2827	2620	2744	2899	2577	2989	2869
Equivalent People	23.0	24.0	25.0	23.0	24.0	23.0	23.5
I.E. Manloading	23.0	22.0	22.5	24.5	22.0	25.0	24.5
Culver City:							
Volume	1716	2044	2186	2126	2123	2694	2513
Equivalent People	21.0	22.0	22.5	21.5	22.0	23.0	21.0
I.E. Manloading	14.0	15.5	16.0	15.5	15.5	18.5	17.5
Monterey Park:							
Volume	2171	2174	2391	2384	2327	2707	2540
Equivalent People	18.0	16.0	16.5	17.0	17.5	17.5	17.5
I.E. Manloading	14.5	14.5	15.0	15.0	14.5	16.0	15.5
Van Nuys:							
Volume	2117	2278	2270	2196	2147	2615	2439
Equivalent People	18.0	15.5	19.5	17.0	18.5	18.5	18.5
I.E. Manloading	14.0	14.5	14.0	14.0	14.0	16.0	15.0
San Diego:							
Volume	3304	3071	3369	3360	3051	3348	3249
Equivalent People	21.5	24.0	24.0	24.5	23.5	23.5	24.5
I.E. Manloading	24.0	23.5	24.5	25.0	23.5	25.5	25.0

Note: Encircled data represents month during which study was conducted.

Figure 10-8. Industrial engineering study, southern region. (Encircled data represent month during which study was conducted.)

WAREHOUSE
VOLUME vs PRODUCTIVITY

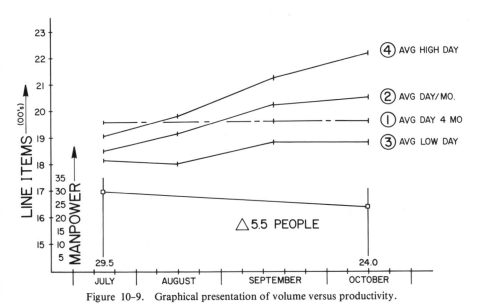

Figure 10-9. Graphical presentation of volume versus productivity.

equate to $15 per hour. On a yearly basis this reduction could be approximately $179,800. Dealing with half a person may cause a problem unless you utilize part-time employees. A reduction of five people would produce an estimated $163,300. Regardless of how the half-person is dealt with, the results are worthwhile.

FOLLOW-UP ROUTINES

What kind of follow-up schedule is required? The follow-up schedule will vary depending on the type of application. Management can review the head count, hours worked, and other productivity measures as the results become available. In most companies this will usually be on a monthly basis. Remember, upper management is interested in the key numbers that show the results of a cost reduction application.

These key numbers will be indicators that the productivity trend is headed in the right direction. Examples of these indicators are:

- Cost per standard hour worked—a change from $26.35 in the previous period to $24.83 in the present period is a positive indicator.
- Cost to process $100 of material out the door.
- Number of line items selected per hour worked per employee.
- Number of orders shipped on schedule.
- A reduction in the number of claims processed and the related cost of each claim.
- Quality of inventory management. This vital measure can be stated in two parts. Part one deals with the integrity of the material in stock versus what the records show. Part two deals with material selected, packed, and shipped to the customer.

From an engineering point of view these items should be reviewed at least on a quarterly basis. Your operation may dictate a different frequency.

A SYSTEM OVERVIEW

Warehouse space requirements can be reduced by high-density storage and a material storage strategy.

Every distribution system can profit from a review of storage modes and a close examination of a material placement strategy. The real question to be asked here is, "How can dealing with the space problem be turned into a cost reduction opportunity?"

Western Electric has spent a considerable amount of time and money in developing answers to the space challenge. A large share of this work has been done by the Warehouse Engineering group located in Ballwin, Missouri. Some of their work has led to many innovative changes in the overall distribution and storage systems.[2] Refer to Figures 10-10 and 10-11.

The first strategy that can be used to minimize space cost is the use of equipment that provides maximum space utilization. High-density storage modes are one answer. However, they may not have applica-

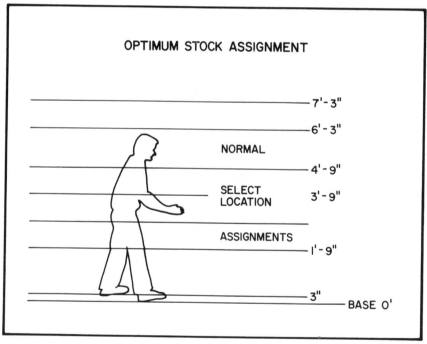

Source: Western Electric Company

Figure 10-10. Productivity improvement tip. Order-picking productivity is affected by where the items are stored on shelving. Remember the 20/80 rule which states that 20 percent of your stock represents 80 percent of the activity? Another rule says that 50 percent of the stock accounts for 95 percent of the volume. So for top productivity, put the most active items in a horizontal band, with the fastest movers where they are most easily reached by the selectors. Check these rules in your operation; you may be surprised. (*Source:* Western Electric Co.)

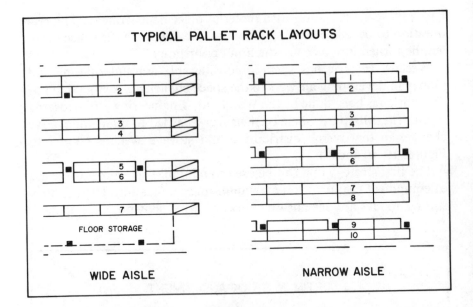

Figure 10–11. Space-saving tip. Sketch of narrow aisle versus wide aisle. Of the 40 percent capacity increase, 20 percent is obtained by narrowing the aisles and 20 percent by compatability with the 25-foot column spacing. Other column spacing can also be dealt with in a similar manner. (*Source:* Western Electric Co.)

tion in every warehouse. Where these modes have application, the following general statements can be made.

- Narrow-aisle warehousing system can provide a 40 percent increase in storage capacity within the same storage area.
- Mezzanines can also provide a 40 percent increase in storage capacity within a given area.
- Tiered shelving can provide a 60 percent increase in storage capacity within the same area.

If your operation could benefit from narrow-aisle warehousing, mezzanine storage, or tiered shelving, your best approach is to sell the concept on the benefits of labor and space.

Conduct your study, inform management of the advantages. Be sure to have the facts at this point. Many cost reductions have been lost due to a small oversight in the selling phase.

Another strategy that will reduce space cost and labor cost is the placement of material. Implementation of this strategy requires a history on each item stocked. From experience it may be a simple task to pick the large runners in any distribution operation. For example, my experiences in the warehouse have taught me that telephone sets are the high-volume items. Adequate storage space must be provided for these items. Also, it is easy to determine the low-volume items. They are usually the ones with a thick cover of dust and last year's inventory label.

The difficult items to deal with are the medium runners. These items may be subject to seasonal demands, fads, advertising campaigns, or any number of other variables. The computer is the key to efficient placement and space assignment.

COMPUTER SPACE PLANNING

Even if your pallet racks and shelving extend from the floor to the ceiling, this is no guarantee that storage cube is being utilized. A computer can be of great assistance in evaluating storage requirements. In this endeavor, like any other, there are two schools of thought. One school says you must include every item in the warehouse. This could mean as many as 16,000 or more for some large industrial operations. The second school says that a selected sample of items is sufficient to determine the required data. Choose the one that fits your needs and budget.

A computer, by sampling a few items, can determine the space needs for the parts you desire to store. The physical measurements of the packaged parts are required for this determination. The activity level of the part is also required, as is the size of storage containers.[3]

Before the advent of the computer in planning or updating a warehouse, everything was accomplished by a manual calculation. This mode may still be adequate today in some operations. If it is adequate, use it. If the computer space planning tool seems to apply, and it will in 99 percent of the warehouse operations, use it.

A computer program can analyze item activity—the frequency at which the part is selected. If you know which parts are selected more than others, you can plan a system and assign storage space based on activity. This will increase selecting productivity.

How to Start

A number of software programs can be evaluated to see which one best fits the need. Your objective is to select the one that can assist you in planning the total storage space. The options are open. Should you purchase the program or call a company to run the program for you?

Certain information will be required regardless of the option you select to analyze space needs in your warehouse.

- A complete description of the number of items to be stocked in the warehouse. In a sampling mode this is required because the computer analysis is based on a sampling of parts—up to 10 percent of the total.
- Measurements of each package size selected for the sample, and the recommended level of inventory. With this input data the computer can calculate the cube required for each item. The output specifies a pallet or shelf-storage mode. The computer will determine the projected number of pallet spaces and shelving units required.

Sample Size and Selection

The input information is your decision. The sample size can be determined in several ways. The sample size will be a function of the number of items that are stocked. A 10 percent sample may be required if the total universe is 5500 parts. As the universe increases in size, the percent can be increased. A data base of 75,000 items may require a sample of 2100 items. This approximates 3 percent of the total.

Selection can be made on a random basis. A random selection can be made on every nth part.[3]

Define the Item and Activity

Each part is identified by a reference number. After identification, each sample is reviewed to determine, the weight, length, width, and depth of the carton to be stocked. This data, coupled with sales data and inventory level, will enable the computer to select the storage mode. Changes in carton size may be a consideration worth investigation. A number of computer simulations may be required to obtain maximum utilization of cube space.

Next, look at the activity level of each item to be included in the sample. Profile information can be fed into the program. This information is required when the storage modes are being determined. Slow runners will probably not have a place in an AS/R system. Once the computer output is obtained, the decision-making tools are available for use.

At this point the ability to interpret output is important. Decisions can now be made that will allow you to allocate space for FAST, MEDIUM, and SLOW items.

Dead items will continue to be a challenge. Look for ways to reduce these from the warehouse. Follow-up in this area will free up valuable space and lower investment.

REFERENCES

1. Kloss, L.K., "Mapping a Distribution Strategy," *Handling and Shipping Management,* November 1980.
2. Western Electric Co., "Space Allocation Manual," April 1980.
3. Nichols, W.O., Jr., "How to Use a Computer to Plan Your Storage System," *Modern Materials Handling,* April 1981.

11
Employees, Expenses, and Expectations

The relationships between declining productivity and inflation surround each of us daily. Inflation has become a constant companion in our daily life. It is easy to identify, but not easy to remedy. The cure requires a concentrated effort from individuals, business, and government. Taking a positive stance to control and then reduce inflation must be our short- and long-term goals. A positive stance to reduce inflation goes beyond periodic campaigns to reduce government spending. A firm action commitment to increase productivity is also required.

How can we develop such a plan? There are no easy answers to this question. However, one thing is certain, reducing input costs through an effective cost reduction program is the first step.

THE PRODUCTIVITY CHALLENGE

To comprehend America's prevailing productivity problem, one needs to review the nation's economic growth pattern. Beginning with the industrial revolution and continuing through two world wars and two decades afterward, the United States seemed to have little trouble chalking up sizable productivity gains year after year. A major factor in earlier productivity growth was an overall shift to nonfarm manufacturing, where output per worker could be dramatically increased with each new mechanization process. American ingenuity demonstrated the country's ability to produce more in less time and with fewer resources, capital, energy, and the human effort and skills needed to manage them.

In comparison, no other nation in the world during that period successfully found the formula for economic productivity—at least not without severe interruptions of wartime losses. But when World War II ended, America's leadership gradually began to be challenged. First came the aid America gave to other countries to actually help them compete with the United States; then, in many subtle but significant ways, restrictions began to confront productivity here which did not apply abroad. Although early warning signals began years ago, it was not until the late 1970s that serious attention began to be directed toward what was happening to U.S. productivity. Suddenly leaders became alarmed that productivity was shrinking. Along with it, the benefits of productivity also began to shrink.

BENEFITS OF PRODUCTIVITY

Productivity improvements offer at least two major benefits: financial gains and job security. To consider the first of these, let us assume that employees receive a 6 percent pay increase even though productivity remains the same. Project this circumstance to most work sites and prices will go up 6 percent, because when productivity stalls while producers' costs go up, there is usually no choice—producers must charge more for goods and services in order to stay in business.

On the other hand, let us say that productivity during this period rises 6 percent; therefore the 6 percent pay raises are justified and prices will not have to be adjusted upward just to cover increased costs. The wage and salary earners thus experience real pay boosts as opposed to "partial inflation" pay hikes.

Job security is also enhanced by productivity. If you work for or own a company that is in business to make a profit, your job is made more secure by improved productivity—this is simple to figure. Even if you are not engaged in a profit-making organization, it is just as important to job security to increase productivity, because underwriters of these ventures are usually taxpayers who are losing patience with bureaucratic waste and are withdrawing support for inefficient institutions and government agencies.

If you have ever heard of Proposition 13, a tax initiative in the state of California, the statement is loud and clear. The taxpayers are tired of the constant increases to feed inefficient local and state operations. This item will be discussed in more detail in this chapter.

A Combined Effort

Contrary to some perceptions, productivity growth does not depend on "harder work." Instead, productivity improves when prudent capital investment increases; when the labor force becomes better trained; when energy costs are kept in check; when research and development results in more efficient technology.

Whether one works in manufacturing, service, or government, we all have a stake in improved productivity. Improved productivity can come from a wide variety of innovations. In an office setting, it may involve better filing, or reporting systems which allow a business to accomplish more during a workday. In a factory, store, or warehouse, it may involve more effective handling of materials, again resulting in more output per workday.

The environment for boosting productivity must be reasonably free of government restrictions and it must also be conducive to capital investment and continued research and development. As voting citizens, we can help make sure the environment follows this pattern which helped make this country a world leader. Support can be given to office seekers who encourage more R&D and capital improvements through tax incentives and fewer legal restrictions that hamper productivity.[1]

MANAGEMENT INVOLVEMENT

Almost every day a national figure expounds a new view of the productivity problem. Most leading business publications directed to members of management dissect the issue and offer certain new and innovative avenues for improvement. Most business leaders will agree that the problem has been building since the mid-1960s.

What is the difficulty in grasping the problem? One reason may be the absence of understanding the true nature of the productivity problem and the continued deteriorating trend. Some leaders in business may say that the problem continues because our efforts are focused on symptoms—not the real problems. The major factor that permits the continued downward trend is the same leading element that must correct it—management.

The problem is truly one that management is entitled to take full credit for. In summary, they must take the problem, and by positive management action, change its course. The only way this can be done is by developing new avenues for cost reduction and improved relationships with employees in all levels of the workforce. [2]

Management involvement is the key that is needed for productivity improvement. Why is this so? Only management, by using the tools at their disposal, can achieve the following.

- Create an employee environment that is conducive to innovation and change.
- Provide by their example an ongoing motivation and involvement program.
- Support new programs that bring all levels of employees into a partial decision-making environment.
- Involve people in generating new ideas. This is essential for management to succeed in their appointed task.

Work Relationships

Some companies, both large and small, do enter into attitude surveys to obtain feedback from employees. Employee perceptions of their company and their jobs are important. Common questions in such a survey may be along these lines:

- How do you feel about top management's involvement in the business?
- Which of the following causes the biggest problem in your work assignment? Rank order each from 1 to 5, with 1 being the highest. (List possible causes.)
- When did your immediate work supervisor last conduct a performance appraisal with you? (The desired answer may be six months to one year. The answer may be a surprise.)
- Have you reviewed your job description in the past 12 months? (A simple question? How would this be answered in your company?)
- Do you feel that your supervisor can be trusted to discuss work-related as well as nonwork-related problems?
- Do you feel comfortable in talking with your supervisor?

- Does your supervisor always want to talk about job-related problem areas?
- Does your supervisor discuss ways for you to prepare yourself for the next job level?
- Does your supervisor hold meetings at least monthly to share group performance trends?
- Do you think you could do more in meeting the company's overall objectives?

Questions of this nature can be asked of all employees on a periodic basis. Sometimes the answers are a shock for management. However, this type of feedback is needed.

Review of Feedback

What kind of feedback can be expected? A simple question indeed. The results of the survey will not be that simple to interpret. Keep in mind that the original intent of the survey was to determine how employees feel about the company and their specific jobs. The specific jobs can range from entry clerical levels to middle management!

In a typical company, the feedback may be along these lines:

- I was not aware that this job had a description!
- My last performance appraisal was just a few words in passing—over two years ago.
- I feel that management could keep employees better informed.
- We don't really know how our department fits into the big picture.
- I like my supervisor but he always wants to talk about the job.
- We are locked into dead-end jobs because of the Equal Opportunity program.
- There is no way to get ahead on this job even though I do have a college education! (This is becoming more common.)
- The company could develop a more effective job rotation program.

The above comments are typical of those from employees at all levels in the labor force. They are not stated to criticize any specific company or management group. They do, however, illustrate the management challenge that must be met as we deal with the productivity problem. As stated before, new ideas are the key to our expanded cost reduction programs. These ideas can come from all employees. Employees who are not in tune with the company objectives cannot be counted on to the same degree as those who feel they are a part of the corporate objective.

THE KEY—FIRST-LINE SUPERVISION

The first-line supervisor is the key in dealing with all employees. The first-line supervisor is the interface between upper management and the employees. In all work operations, including the production shop, the warehouse, or the office, the first-level supervision is the key agent to the introduction of change. Several considerations for the first-line supervisor are:

•Review the job description for each employee's assignment. Make sure the employee knows what has been covered. Discuss the key parts of the job assignment. Ask the employee to share questions and concerns.

•State what is expected of each employee. Show how each individual is needed to insure that objectives are being met. Explain how the effort of the group and each individual is needed to insure that objectives are met for the total work unit.

•Review how the job is currently being measured. A shop job is usually measured in terms of efficiency. As an example, if an employee takes one hour to manufacture a part with a predetermined standard time of one hour, the efficiency is 100 percent. (1.0 hour/1.0 hour = 100 percent). However, if it took 2.0 hours to complete the 1.0-hour assignment, the efficiency would be 50 percent. Keep in mind that the productivity is not solely a matter of the equipment utilized but how effectively it is used.

•Office jobs will require a different type of measurement. A typing pool may measure output as the number of typed pages per shift—modified by a difficulty factor. Other jobs that deal in a service mode, such as an order desk in a large department store, may use the number of complaints handled compared to the number of those successfully handled. A total of 32 complaints were handled in an eight-hour shift; 24 complaints or 75 percent were handled without the assistance of the department manager. Is this a good record? That depends on the training that has been provided versus what is expected from the average trained employee.[2]

Report of Monthly Accomplishments

One time-tested approach used by many first-line supervisors is the monthly letter to upper management. The report can be a brief, concise, yet informative piece of information that comes in advance of the formal statement of performance that is published for upper management.

A brief handwritten letter of monthly accomplishments can be prepared in less than 15 minutes. In general terms it could include a brief statement on the following items of interest:

- Efficiency for the current period against the forecast.
- A statement of quality as measured against an expected level.
- A statement of in-process material, the amount of work from the start of the process to the end of the line.
- Any new ideas that may lead to cost reduction in the coming months.
- A brief statement of any problem areas that were encountered in the month, and action steps taken to alleviate the problems.

HOW TO BECOME A PRODUCTIVITY SUPERVISOR

What can you do to become a productivity supervisor as opposed to a traditional supervisor? This is a question with a wide range of answers and the answers may change with different industry segments. However, the following ideas are suggested if you want to become a productivity supervisor.

- Define the job to be done and standards of performance.
- Encourage employees to submit ideas that will improve the job.
- Ask employees for their opinion on job-related items—remember, they do the job for eight hours daily.
- Try to explain why things are done in a certain way when employees question them—it may be due to corporate policy or some other local instruction. Perhaps now is the time to update the existing method.

- Acknowledge a job well done, perhaps a monthly quality award for the employee with the best quality record, or a yearly award for perfect attendance. List several others that may apply.
- Keep an open mind for methods and ideas that may work in today's operation.
- Squelch rumors and deal in facts. Nothing can cause unrest quicker than a constant flow of rumors. Needless to say, this adds little to the quest for increased productivity. The supervisor's best approach is to answer each question to the best of his or her ability—or just to say, "I don't know but will try to find the answer."

Effective Management Is the Key

Management must generate an environment that fosters productive change. Innovation and superior performance must be recognized and rewarded. Employees must be able to perceive productive change as a road to increased monetary returns. Many employees perceive the opposite. Some employees may recall that those who tried to initiate change were viewed as malcontents and troublemakers. These memories do little to foster new ventures into areas that have already been proven to be bad experiences. A new idea, a change of method, a different system involves the possibility of failure as well as success.

The absence of management motivation will tend to keep employees in the status quo mode. When this occurs a terminal case of mental rigor mortis takes place. A major commitment by management to create a positive climate is a first step. In today's environment there is no room for managers who choose the warm feeling that apathy tends to generate.

Starting an effective cost reduction program can be a positive step into solving the productivity problem. As a member of management your support is essential. Technical problems can in general be solved. People-related problems are more delicate and require more effort for the new programs to succeed.

THE BIG PICTURE

A cost reduction program can be designed to fit the needs of any company. Each company, regardless of size, has the same areas of opportunity: labor, material, capital, energy, and other related expense. Every company will present a challenge since the percentage of each item will differ.

Some industries will present a challenge in dealing with problems associated with labor and capital. Others will be heavy in energy requirements. Material and related expenses will also present avenues for cost reduction. In order to secure a feel for your operation you will need to break out the percentage of each of the above items that make up the cost of sales.

Today, in business operations we tend to look at big as being better. This may not always be true. If you view sales as a measure of bigness, let us take a look at some data.

Fortune magazine presents a yearly summary that contains the 500 largest U.S. industrial corporations. A review of the *Fortune* 500 list makes interesting reading. The list is keyed to sales in the current period. Sales are of vital interest in every company. The profits resulting from sales are in fact the real measure of how effective management has been in running the company in the period under observation. The hidden item that is not apparent is how well has management controlled costs. A bright profit picture in some companies may be as hard to explain as a dollar loss is in other companies. The lack of firm cost control programs may or may not be evident in any specific period under review.

Many partial productivity measures can be used to compare one company against another. Some of the most common partial measures are:

- Total return to investors: This includes dividend yield and price appreciation to an investor. In 1980 the median for the *Fortune* 500 was 21.05 percent. How does this compare with your company? This information may not be easy to obtain if your company is part of a large conglomerate.
- Assets per employee: This is the total value of assets at year-end divided by the number of employees. It provides an insight into the nature of the company's business. Is the operation labor or capital intensive? Assets per employee in 1980 for the 500 largest companies was $55,505 per employee.
- Dollar sales per employee: This is obtained by dividing total sales for the year by the number of employees. The median sales per employee of the 500 listing in 1980 was $78,147 per employee.

**Table 11-1.   Partial Productivity Data
(dollar sales per employee vs. assets per employee).**

RANK	COMPANY	SALES	ASSETS	RATIO
1	Exxon	$583,998	$320,338	1.823
3	General Motors	$ 77,384	$ 46,355	1.669
8	IBM	$ 76,808	$ 78,243	0.981
10	General Electric	$ 62,087	$ 46,047	1.348
13	ITT	$ 53,246	$ 44,302	1.201
19	U.S. Steel	$ 83,742	$ 78,752	1.063
22	Western Electric	$ 67,978	$ 45,466	1.495
30	Boeing	$ 88,675	$ 55,797	1.589
35	R. J. Reynolds	$101,286	$ 88,175	1.149
307	National Semiconductor	$ 24,349	$ 13,942	1.764
	All-industry average (includes 500 companies)	$ 78,147	$ 55,505	1.408

Refer to Table 11-1 for a listing of relationships among a selected sample of companies. Note the variance in dollar sales per employee and assets per employee in this sample. Compare these with the all industry data. These numbers are interesting to compare but may not reveal the cost reduction potential that is masked under the daily operations of these, the biggest and most successful companies operating in the United States today.[3]

The Final Question

How to improve the operations is the daily challenge for upper management. One thing is for sure, some management segments do not stay around long enough to plan for the long-range program—or should it be stated in the reverse manner?

In the final analysis the profit motive is the prime mover or prime remover of upper management when the profits are not achieved. Increased sales are always a goal. There are trade-offs between savings in the current operation versus increased sales. In Western Electric and other companies, $1 saved is worth $20 in sales. Why is this concept important? A company may be able to achieve the same impact on bottom-line results by saving $1.0 million through an effective cost reduction program versus generating new sales of $20 million.[4]

Stated another way, the $295 million saved by Western Electric in 1980 would equate to additional sales of $5900 billion. Regardless of how cost reduction is viewed, it must be utilized as a tool for survival.

As a topic for the annual report, cost reduction is being mentioned more than ever before. If you have any doubt, scan a few annual reports at random. You may be surprised.

EXPENSE CONTROL

Let us review several tried and proven items that illustrate effective expense control.

1. Revised Lighting System

Pacific Southwest Airlines, with headquarters in San Diego, installed a new lighting system in its maintenance hangar. A cost reduction study showed that the yearly cost of energy could be cut approximately $48,000.

The study was based on 3650 hours per year at $.10 per kilowatt-hour. Lighting levels were raised 50 percent and the payback period was estimated at five months based on current-level savings. The initial plan called for relamping a 64,000-square-foot hangar with 288 new 400-watt mercury-vapor lamps. The initial goal was to provide 75 footcandles at the floor level, and 90 footcandles on the top of 35-foot work stands.

A meeting with local utility officials convinced PSA to utilize 144 high-pressure sodium lamps versus the 288 mercury lamps first planned. Cost of the change was estimated to be $20,000. The new system is mounted 55 feet above the floor on 22-foot centers.

In summary, a cost reduction of $48,000 yearly was achieved at no sacrifice to the work environment.[5]

2. Energy Management

An energy management system was installed in a 60,000-square-foot office building. Despite three rate increases, the system has resulted in a 10-percent drop in energy costs.

The system was installed in November 1980. In the first five months of operation, electric consumption was reduced about 15 percent, and the demand level about 20 percent. Total electric bills were reduced an average of $1560 per month.

The system utilizes time-of-day, duty cycling, and demand control features. Savings were derived mainly from the air conditioning system. The building utilizes 67 four-ton heat pumps. Formerly, the system operated constantly. In the new mode it operates only when needed.

Building specifications are incorporated into the control system. Certain parts of the building are open only from 8 A.M. to 5 P.M., but other offices are open on a random basis. The system is programmed to handle these different demands. Payback for the system is estimated to be nine months. Return on investment is estimated to be a hefty 136 percent.

The system was supplied by Honeywell, Inc.[6]

3. Control of Telephone Costs

All expense aspects of the daily business should be reviewed for cost reduction potential. Like any other expense item, the telephone is a candidate for close examination. History has indicated that the cost increases associated with telephone usage have been well below other operational elements such as labor, material, and energy.

Can savings still be obtained from what already is a known bargain? The answer to this question, I believe, is yes!

In most business operations, both large and small, the telephone is always there. This results in the telephone being taken for granted. This is not true in all cases. Substantial interest is noted when a new system is being considered for upgrading the existing service.

Take a good look at your present telephone service network. The first question to ask is, "Does the system really meet the needs of the business?" In order to secure a completely valid answer it may be necessary to call in an outside consultant. Even before the consultant is called in, the following questions can be answered. These questions will provide a starting place in the audit process.

- How many employees share the same local line?
- How many employees are provided a speaker phone?
- How many card dialers are being utilized?
- How many key sets or multiple line sets are being used?
- Are the telephone sets equipped with a touch tone or rotary dial?

These are basic questions that can lead to real savings when explored.

Another consideration not to be overlooked is the network of services that link the local operation to customers in other parts of the country. Evaluate the full range of services that are available. Some telephone consultants may even be willing to work on a shared savings basis without any up-front payments for their specialized services. One large Bay Area electronics company did such a study and saved $275,000 the first year.

4. Invoice Processing

The most effective cost reduction involvement is to prevent the expense from occurring. Take a good look at the method of processing supplier invoices in your operation. Examine the number of invoices that offer an inducement for early payment such as the 2 percent for payment in 10 days.

Analyze a typical month and keep track of the average time and flow of the invoices through the typical accounting maze. Consider the following situation:

A company pays on the average 6000 invoices a month. A review of the invoices indicates that 80 percent offer the typical 2-percent discount for payment in 10 days. Average monthly payments, that offer discounts, approximate $960,000, or the typical invoice has a value of $200. In this case the savings would equate to $4.00 per invoice. A capture rate of 100 percent would equate to potential savings monthly of $19,200. In reality it is unlikely that a 100-percent payment rate can be achieved.

The important thing is to establish a goal in this area. One key question to ask is, "What level of discounts have been obtained in the past?" In some organizations this data may be hard to generate. After your investigation has been completed, establish your goal for obtaining discount payments.

A starting goal could be to obtain 90 percent of the discounts. This would provide a monthly savings of over $17,000 and yearly potential in excess of $200,000. Programs of this nature that have been in operation for one or more years find that goals in the range of 94–97 percent are possible to achieve. The important thing is to develop your goal and place the plan in operation.

5. Time Paid—Not Worked

As stated, the productivity problem relates to inputs and outputs. Numerous plans have been implemented by companies large and small that deal with cost input containment and the expansion of various types of output. One area that may offer substantial cost reduction is an examination of dollars spent in the area that is known as time paid—not worked.

To be more specific, this deals with dollars of expense that are built into the various overhead recovery rates. These various cost inputs include vacation time, holidays, personal time off, sickness, and time lost to many different kinds of accidents that may or may not be job related, and last, special holidays or days away from the job that have been won through union contracts or other means. Each of these areas may involve the expenditure of dollars in a needless manner. These areas may in fact represent untapped cost reduction that can be easily obtained with just a little hard work.

The first step is to develop a profile for the organization by department for days off the job and the related dollar impact. This exercise will be conducted using your present guidelines for payment of time not worked. One of the most fruitful areas may be a close examination of personal days off the job that compare the days paid versus non-paid days and the average days lost per employee.

The profile data sheet when completed may look like this.

GROUP	NO. PEOPLE	NO. DAYS PAID	NO. DAYS UNPAID	TOTAL DAYS	AVG. FOR EMPLOYEE
Management	30	96	24	120	4.0
Engineering	23	115	28	143	6.2
Accounting	30	168	98	266	8.9
Marketing	16	64	16	80	5.0
Production	350	2858	850	3708	10.6
Warehouse	75	690	240	930	12.4
Composite	524	3991	1256	5247	10.0

Once the profile is completed, the real analysis can begin. Each group can be analyzed for trends against last year's results. In this example the ratio of paid days is 76 percent of the total. When converted into costs on a yearly basis, the dollars can be very large. When the number of paid days is converted to dollars in a typical company, the yearly cost would approximate $350,000. A 20-percent reduction across the board for all employees in the days paid category could produce a cost reduction savings in the area of $70,000 yearly. This is only the beginning. A reduction in the overall absence rate per employee will have many other side benefits such as quality, morale, and a more stable labor force.

As management's program is introduced it will probably include the following:

- A review of each employee's record for the last three years.
- A well-planned information program to inform all employees of the reasons for the new emphasis in control of time lost from the job.
- An action plan that is firm but fair to all employees who do not meet acceptable standards.
- Weekly conference periods between problem employees and their immediate manager to review improvement or the lack of improvement.

In summary, this firm personal approach will work and produce the desired results when it is administered by a prudent management team.

People Considerations

A number of items with cost reduction potential have been discussed. In our quest for increased productivity, people and increased involvement will be the elements that can turn the tide.

White-collar workers—people in clerical, sales, professional, and managerial jobs—now comprise more than 50 percent of the U.S. work force. There is a general feeling that productivity of these workers is declining even faster than that of blue-collar workers. There are many views that may or may not apply to the current problem.

- Some may say that technology systems can not keep pace with changing demands, the result—increased paper shuffling.
- Some may say the younger generation is harder to work into the average job.
- Some would say that today people are less willing to dedicate their total lives to the company.
- Some others would say that declining productivity is truly the fault of ineffective management.

There may be some truth in each of the above comments. Let us take a look at some aspects that relate to clerical productivity.

In his paper, "Don't Overlook Clerical Productivity," R. K. Martin shares a number of unique observations. The paper analyzes the challenge from different viewpoints.

One of the secrets to successfully applying work measurement to a clerical staff is to let the subjects get involved. Here is a simple, straightforward program that any organization can use to improve clerical and managerial performance.

R. Keith Martin
Baruch College City University of New York

DON'T OVERLOOK CLERICAL PRODUCTIVITY![*]

The quality of supervision is directly reflected in the performance and productivity of the clerical staff responsible to that supervision. There are several techniques which, when properly used, provide a rather exact measurement of clerical efficiency and supervisory or managerial effectiveness. These techniques, four in number, are work sampling, work measurement, work simplification, and work scheduling.

The techniques both measure the performance of an organization and indicate the areas in and means by which the effectiveness of the organization can be improved. Generally, improvement in effectiveness can be obtained in one or more of the following ways:

- By making the big change—in organization structure, in policies, in functions, and procedures,
- By making the many small changes which, while individually are seemingly insignificant, have a tremendous cumulative impact,
- By stimulating the employee's desire to do a full day's work, and
- By scheduling work in such a way that each employee is given one hour's worth of work for each hour he or she is paid to work.

The Employer's View

We will examine a program that is extremely effective in the last three sources of improvement. This program is unique in two ways: the manner in which the four techniques of work sampling, measurement, simplification, and scheduling are integrated to bring about the desired results; and the relatively low degree of accuracy implicit in the approach to work measurement.

It is through measurement in one form or another that management learns what it receives from its employees in return for its payroll dollar. And it is from criteria called performance standards, which are developed from the results of measurement, that management can establish within various limits what it should receive in return for its payroll dollar. Accuracy in this area means the preciseness of measurement of human effort.

The matters of accuracy of measurement and of standards of performance were carefully considered. During the past fifteen years the need for a high degree of accuracy has been stressed by a number of professionals who design and install work measurement systems as well as by most users of work measurement techniques. However, the principal reasons why more companies have not adopted work measurement as a normal management control technique appear to be the relatively high cost of installing a system of precise measurement, the complexity of such a system, and the high cost of maintaining such a system after it has been installed.

When reviewing the results achieved by companies using the various work measurement techniques available today, it becomes apparent that great precision in measurement is not necessary to insure attainment of the majority of the potential benefits. Work measurement and performance standards are means to an end, not an end in themselves, as complex systems tend to become. Work measurement and performance standards are merely techniques by which definite yet reasonable goals may be reached. The definition of the goal should, in each instance, determine the technique to be used. Thus, the greater the flexibility of a system of measurement, the broader is its range of applicability within the business community.

It should be apparent that no one type of system can provide the scope and flexibility to serve effectively a range of companies differing not only in size but in the variety of objectives sought. An engineering firm employing five clerical people may want to determine whether or not a bookkeeping machine would pay for itself through man-hour savings; a manufacturing company may be concerned over why its data processing department constantly requires overtime; or a utility company may want to analyze the feasibility of reducing the cost of its clerical payroll by 20%.

Another factor which influenced the development of the program was the opinion that about one-half of the usefulness of work measurement lies in its motivating influence upon employees and the balance of its usefulness is as a means for discovering and identifying the conditions which militate against efficiency. As soon as the circumstances which cause waste motion and lost time are revealed and described, they can be corrected through work simplification—methods improvement, form design, office layout, etc.—the establishment of a reasonable work pace, and work scheduling.

The Basic Program

The program consists of seven separate but related parts.

1. A procedure survey: To define the organization, describe the procedures in general terms, and determine the major tasks of each clerical position to be analyzed by work sampling and work volume measurement.

2. Work sampling: To determine the time required to perform each major task, the time spent on nonproductive work and the amount of idle time. The work sampling observations are usually made by company employees—supervisors or others trained to do the job.

3. Work volume measurement: To determine the volume of work produced in each major task category. The data required are recorded by each employee.

4. Work simplification recommendations: For the application of whichever of the techniques of scientific management are necessary to eliminate idle time and wasted effort. For example: improved organization, better forms design, improved mechanization or demechanization, better office layout or work station layout, procedure changes, quality control, and supervisory training.

5. Work scheduling: To provide supervisors with the techniques and training which will enable them to give each employee one hour's worth of work for each working hour in the day.

6. Comparative reports: To inform management each month of the manhours required by function or section and of the volume of work produced during these hours. The reports may include the unit time or the number of employees required to produce specific types of work. These reports are prepared by the supervisors.

7. Engineered performance standards: Developed for specific high volume, repetitive operations in order to " take a reading" of the efficiency or general work pace of a position or section. The standard hours are compared with the actual hours to determine the efficiency of the balance of the operations of the position or section by interference. Engineered standards of per-

formance are developed from predetermined time values in cooperation with the supervisors.

First, a Procedure Survey

The main purposes of the initial procedure survey of a clerical section are to define its organization in functional terms, to describe the procedures in writing, and to determine the major clerical tasks for which time requirements will be determined by work sampling and volumes developed by work volume measurement.

During the course of the survey it is essential that the person directing the study, whether a company employee or an independent specialist in the field, win the confidence of the personnel involved, because he will soon become the exponent of new ideas and changes. The most satisfactory and effective way to bring about change is by persuasion. If the analyst (as we shall call him) establishes a reputation for open-mindedness, patience, and independence during the fact-finding phase, persuasion will be much easier later on. The adage "never underestimate the dignity and importance of the human ego" should be carefully observed throughout the course of the study to minimize the possibility of personnel and morale problems.

While making the survey and getting the facts, the analyst should consciously try to evaluate various phases of the section's operations including general efficiency, organization, and communications, procedures, layout, forms, management controls, ability of first-line supervisors (determining whether they are group leaders or real supervisors), turnover, training, planning, scheduling, personnel policies, and operating policies. He should identify the key supervisors—those who are most likely to influence management decisions. He should also identify the chief operating problems and deficiencies by observation and by discussion with the supervisors.

The survey itself should start with a chart of organization in which functional titles are used. This should be followed by a written general description of the procedures, estimates of the volumes, and estimates (if possible) of the man-hours involved in the different tasks. This information should be obtained from the first-line supervisors, preferably at their desks.

During the survey it is important to establish a benchmark of present efficiency against which to measure future improvements. This need be no more than agreement with the supervisors on the average month's volume of work and the man-hours required to produce it. The analyst should maintain a list of suggestions for improving operations, whether self-originated or received from others, because it is desirable to be prepared to make some recommendations for improvements, as occasions may offer, as quickly as is practical.

Then, Work Sampling

Work sampling is used to accomplish two purposes in this approach to clerical cost control. Its primary purpose, as a fact-finding technique, is to determine the amount of time devoted to the principal productive activities and the amount of time consumed by the principal nonproductive activities. Its secondary purpose, as a training aid, is to stimulate creative thinking and interest in scientific management techniques in the minds of the supervisors.

The basis of work sampling involves actual observation of clerical activities at various time during the workday. The activities observed are posted to an observation sheet by the observer. The work elements or activities as determined by the procedure survey should be listed across the top of the page. A different sheet is required, therefore, for each different type of clerical function such as payroll clerk, accounts receivable clerk, receptionist, messenger, etc. The names of the employees within each job position should be listed down the side of the sheet. Observations are indicated by "tally-marks."

Work sampling is an effective aid in training supervisors. Experience indicates that when supervisors are required to make the observations, after proper indoctrination, they see their sections in a different light. They are forced to see for the first time many things they have been looking at for a long time. As a result of this, they come up with sound ideas for improving operations. Also, acceptance of the findings by supervisors is definitely stimulated when they make the observations themselves.

Sampling of clerical activity should normally extend over at least a two-month period. This may seem like an excessive amount of time, but if the study is to be meaningful, adequate time should be taken. An analysis of the methods and results of many studies indicates that, generally, two months is required.

During the first week of the study, only one observation cycle should be scheduled per hour. At the end of the first week, the supervisor will have learned most of the element codes and will generally be more proficient. One observation cycle per hour will provide enough observations after one week to permit a review of the elements. It may be necessary to eliminate some, or it may be necessary to add some. After the first week, two or even three cycles may be scheduled per hour depending on the work load of the supervisor and the requirements and characteristics of the study.

One observer should not be required to observe more than about 15 people. If situations are found where a supervisor has more than 15 clerks or where a number of the clerks are out of sight of the supervisor, it is best to assign some of the observation work to a group leader. If this is impractical, the observations of the individuals should be randomized.

It is necessary to schedule an observation cycle during each hour to insure homogeneity throughout the day, in view of the fact that the entire study will be limited to a few thousand observations. The minute during the hour when the observation is to be taken should be randomly selected. Tables of random numbers are available for use in making the selections.

To determine appropriate work elements, reference should be made to the general survey. The main tasks performed by each employee should be selected. The same elements will be used on daily production report sheets which will be prepared by each employee. Therefore, it is important that only those tasks which are readily measurable in terms of volume of work be selected as major work elements. "No work" elements such as walking, talking, waiting for work, etc., should be included in the observations.

Where an extended workday is encountered due to regular overtime or a split or second shift operation, additional observation cycles should be scheduled. It is particularly appropriate to measure the extended hours of work because the efficiency of clerical operations tends to drop off during the afterhours work periods.

The indoctrination of the supervisor in work measurement should begin during the general survey when his interest and curiosity should be aroused. The analyst should be enthusiastic about work sampling and work volume measurement and should do his best to convince the supervisor that these are valuable tools. Training in how to make work sampling observations can usually be limited to instructions. The main points to be emphasized are:

1. Adherence to the random times to start the observation cycles,
2. Entering as the observation what the employee is actually doing at the moment—not what he was doing a moment ago or is obviously about to do next. This point is vital to the success of the study and cannot be over-emphasized.

If it is found, after two weeks, that more than 50% of the major elements occur less than 5% of the time, too fine a breakdown of elements has been made. Some of the elements should be consolidated. If a consolidation is not practical or desirable, there are two alternatives. One is to prolong the study beyond its planned length and thereby get more observations. The other is to determine the man-hours involved in the small tasks by using engineered time standards.

As the work sampling study progresses, deficiencies in the operating procedures, organization, layout, forms, etc., may become evident. These should be corrected as rapidly as possible by agreement between the supervisor, management, and the analyst. All recommendations leading to improvements, whether accepted or not, should be documented.

Work Volume Measurement

Work volume measurement, like work sampling, is used to accomplish two purposes in this approach to manpower management. Its primary purpose, as a fact-finding technique, is to determine the volumes of work produced by individual employees and, by addition, the volume of work produced by organizational units. Its secondary purpose is to stimulate employees to do a better job.

Under this technique each employee is required to record daily the number of principal units of work he produces. He does not record, nor is concerned with, the time required to produce the work. Determination of the principal units of work to be measured is made during the initial procedure survey. Principal units of work are identical to the major work elements used in the work sampling phase of this approach.

The accuracy of work volume measurement applied in this manner has been well established through many years of experience and use. Occasionally, carelessness or deliberate magnification result in distorted reporting at the outset. This becomes evident, even though not suspected, when the reports are reviewed at the end of the first or second week, because work moves from one position to another and from one section to another. Thus, the output from one position or section must correspond quite closely to the input of the succeeding position or section. It is only necessary to have the supervisor of the offending section point out to his group as a whole that they apparently processed more than they received (or passed on) to produce accurate reporting.

The psychological effect on the employee often results in a substantial improvement in morale and nearly always in an increase in production despite the almost universal initial apprehension of supervisors and managers that it will have the opposite effect. Morale improves because this type of measurement apears to the employees to guarantee them recognition by management of their efforts. Furthermore, recognition is based on what they say they do. Response to trust is a human characteristic. Production increases (wherever it has been below standard) due to pride of accomplishment when accomplishment is recognized. Also, this type of measurement, in the eyes of the employees, seems to promise greater equity in the distribution of the work load and clearer identification by management of deserving conscientious workers and undeserving, inefficient, lazy, or incompetent workers. And, it does.

Work volume measurement usually should be installed as a permanent control technique and remain in operation indefinitely. The production data collected each week form the basis for the management control reports with which management may exercise control over the size of the indirect labor force and thus the payroll cost.

A daily production report form should be used. Initially at least, a separate report form should be prepared for each position. The name of the position, division, and department, along with a description of the units of measurement (major work elements), should be included on the form.

The installation of work volume measurement should be preceded by a talk to the employees by a top official. He should explain what the employees are being asked to do and should tell them why. Among other things, he should (unless impossible) guarantee that no one will be fired as a result of the program. He should explain that any staff reductions will be accomplished through normal turnover and that transfers of individuals may be made, after consultation with the individuals involved. He should state that management is basically interested in knowing what it is getting from its employees in return for what it is paying; that it is interested in establishing an equitable distribution of the work load; that it wants to establish a factual basis for recognition of individual merit; and that it wants to find out what the unit costs are of the various types of work. After the talk to the group the supervisor should offer to assist anyone who feels he has a problem in deciding the best method for measuring his work. Work may be measured by using a ruler to count punched cards, plain counting, etc.

During the first three days the supervisor should be prepared to answer questions raised by individuals. Many people will fear that they are not doing the job correctly and will want reassurance. Many people will feel that the elements selected for them do not adequately reflect the scope of their jobs (which each person rightly feels is highly important). A convenient way to handle this is to agree and be persuaded that more elements should be included. Insist that the employee record the work for the elements as shown on his sheet, but suggest that he write in any additional elements he feels strongly about. It is possible that some additional significant elements will be discovered this way. If so, they should be incorporated in both the work sampling and work volume measurement forms. After two or three weeks, employees tend to lose interest in writing down their own extra elements and gradually stop. It often happens, after the first month's report has been analyzed, that some changes must be made on the daily report forms.

Work Simplification

The principles of motion economy and the conservation of energy are applicable to all tasks performed by people everywhere. The office is no exception, nor is any part of the office an exception.

To be a successful simplifier of clerical work, a person must train himself to be motion conscious, and he must become familiar with the simpler machines and devices which are available for office use. He must be inquisitive and he must learn how to ask "why," not in a challenging manner, but candidly.

It is a rare experience to come across an operation which cannot be simplified at least a little. It is important, however, that a person be selective in his efforts at work simplification. It is important to consider only those jobs which involve the largest number of man-hours per month. For example, in an inventory record procedure, each posting operation may take two minutes of a clerk's time, and each month-end reconciliation may take two days. It may be easy to see ways of simplifying the reconciling operation and not so easy to see how to simplify the posting job. But if 2,000–3,000 postings must be made each month, the analyst should tend to disregard the former and concentrate on the latter.

Both the work sampling observations, which reveal the man-hours involved in each major task, and the work volume measurement reports, which show the number of times each task is performed, are excellent guides in determining productive areas for work simplification. Additional clues include the following:

- Excessive walking indicates a need for simplification of the work flow and possibly a more efficient floor plan layout. Only messengers are working while walking.
- Excessive talking may indicate a need for a better floor plan arrangement, partitions, or possibly sight supervision.
- Many private stenographers or a large volume of dictation often point the way to savings through the use of dictating machines and typing pools.
- A heavy copying or retranscription volume may indicate the need for photocopying equipment, more carbon copies, or duplicating equipment.
- Many filing hours may call for a reorganization of the records or a central file system.
- Peak loads, high unit costs, low individual productivity, unequal individual output within positions, and idle machines may all be clear signs that scheduling is needed.
- Low typing or key punching production may lead to forms analysis and the redesign of some forms.
- A high volume of activity in unit record files such as punched cards may lead to savings through the acquisition of electrically operated files.
- High volume jobs in tabulating departments often indicate the need for machine load records, for putting wheels on card file cabinets, or for assigning one man to two or three tabulators and assigning others the task of transporting cards to and from the machines.

- Poor housekeeping, as evidenced by dust on the furniture; waste paper on the floor; magazines, files, records, and paper piled on tables, desks, and tops of file cabinets; disordered file drawers; and records, papers, and documents in desk drawers, is usually a good indication of poor quality work, reduced productivity, and a feeling of indifference among the clerical force.
- Poor work station layouts—where frequently used materials, supplies, machines, and reference material are not within reach—cause waste of time and effort which often can be eliminated through a sensible rearrangement of the work station based on the principles of motion economy.

Work Scheduling

Work scheduling is the day-to-day and hour-to-hour responsibility of the first-line supervisor. Foremen in charge of direct labor workers usually understand this responsibility clearly because the essence of a foreman's job is to see that production is carried on efficiently. Supervisors of indirect labor usually understand this only vaguely, if at all. The reason for this is that management has not been as concerned with the productivity of indirect labor as that of direct labor. As a result, clerical supervisors are rarely given staffing goals. They almost never receive training in how to measure production or how to develop and use production standards.

It is, in fact, too often true that supervisors of indirect labor are not really supervisors. Instead they are lead clerks, or group leaders with the title of supervisor. One purpose of the program is to make them supervisors, at least to the point where they use the production standards developed for them, schedule and distribute all work to each employee, record the time required by each employee to complete each batch of work, and prepare at least one monthly report for management on the efficiency of their sections.

To be able to assign an employee an hour's worth of work, the supervisor must know quite accurately the time required to produce a unit of work. This information, developed during the work volume measurement and work sampling steps of the program, is available on the unit time calculation sheet for the last week the work sampling observations were made. The unit times on this form are pure work times; they do not include an allowance for personal need and fatigue. Before they may be used for scheduling purposes, each must be increased by one-sixth, which is an acceptable allowance for personal need and fatigue.

The question will inevitably come up: Should employees be shown the unit times or the production "standards"? The answer is that usually it is not possible to keep them a secret. And further, they are not real production "standards." More accurately, they are rules of thumb for the supervisor. A strong argument can be made that "in the interest of fair play" the employee should know what is expected of him.

And Finally, Some Controls

Management controls are the reports prepared for management which show the relationship between the volume of work produced and the number of people or man-hours required to produce it. They show the performance levels attained by employees relative to the performance goals established through work measurement. They also show the effectiveness of first-line supervisors, who are the members of the management group directly responsible for the productivity of the employees.

The establishment of a relatively simple system of control reports is an essential part of this program. Without such reports it is difficult, if not impossible, for the company to realize the savings potential inherent in this program. The purpose is to present the facts on production and personnel requirements to management periodically in such a way that management may be kept well informed on the performance of each section and may be able to react quickly and intelligently in the event the performance of any section begins to decline.

One report should set forth clearly and concisely the payroll reductions resulting from this manpower management program. It must be prepared on each assignment and presented personally to top management. It is often called a "Staff Required" report. This report gives management a specific staffing goal and pinpoints the areas where savings can be made.

In the event that "standards" developed in a branch office are to be applied in other branch offices (and this should occur only where identical procedures and forms are used), the volume figures for the other offices should be obtained and multiplied by the unit time data developed in the pilot office to determine the staffing requirements for each of the branches.

In the case of small offices or small sections in which fewer than ten employees perform a large number of distinct tasks—more than four per employee—it may be desirable to develop and use volume-time factors instead of actual unit times in calculating "standard" hours for reporting purposes. The use of volume-time factors reduces the number of work output items which must be counted or measured and reported each month. Volume-time factors are developed by selecting several key tasks, then adding to the total productive time spent on these tasks, the total productive time spent on all other tasks where work volumes vary directly with the key tasks. In other words, add the total productive hours spent on the other tasks to the total productive time spent on the key task. Calculate and add the two allowances (personal need and fatigue), then divide the total time by the volume for the key task. The quotient will be a factor which, when multiplied by the actual monthly volume for the key task, will give the total man-hours required for all of the tasks included in the factor. When volume-time factors are used to simplify the reporting of volumes, it is usually necessary to use one or more constant factors in order to account for the time required by tasks which are unaffected by volume changes.

A periodic control report is essential. Management should receive it monthly or every four weeks. This report is prepared by calculating the standard number of hours or people required each month. In the report, the standard hours or employees are compared to the actual hours or employees and the variance is expressed as a percentage of the standard and may be called "efficiency" or "attained performance level."

As stated earlier, these two reports, or their equivalents, must be prepared as part of the program. The selection of the report format and the detail to be included should be determined during consultation with the executive for whom it is intended.

These reports are simple and basic. The comparative reports make possible "management by exception." They highlight supervisory performance as well as report employee performance, but if they are to be of significance to an organization, management must insist upon receiving them, must read them, must ask "why" when productivity decreases or man-hours increase, and, last but not least, accord appropriate recognition to supervisors and employees where man-hours decrease and productivity increases.

REFERENCES

1. American Institute of Industrial Engineers, "Productivity Is Everybody's Business," September 1981.
2. Young, S.L., "What's Needed for Productivity," *Production Engineering,* March 1981.
3. Worthy, F.S., "The 500," *Fortune* magazine, May 4, 1981.
4. Western Electric Co., "Where Do I Go From Here," *Newsbriefs Background,* November 20, 1978.
5. "Lighting System Cuts Cost," *Industrial Maintenance and Plant Operation,* September 1981.
6. Ibid.

12
Public Sector Cost Reduction

There have been many misconceptions about the application of proven cost reduction techniques in various levels of the public sector. For years all segments of government—local, state, and federal—have been engaged in runaway spending campaigns. How has this become possible? Simply stated, the governing group has continued the increased expense outlay. In order to help close the gap more money had to be raised to control the deficit pattern—not to eliminate the deficit.

AN UNLIMITED SUPPLY

Until the late 1970s most of our elected officials at all levels appeared to be "free spenders." A free spender can be defined as one that continues to expand the outward flow of goods and services beyond that of projected income.

This is only one part of the problem. Keep one critical fact in mind: that each dollar expended by our elected officials has to be raised from some input source. The biggest share of these input dollars comes from taxpayers in all walks of life.

There is no unlimited source of tax dollars to feed inefficient government operations. People in all sections of the country are beginning to speak out on this issue. How effectively are the people speaking? This depends on the section of the country that you reside in. The loudest outcry from the people developed in the state of California. This outcry was in the form of Proposition 13, a tax relief initiative that was voted in by more than a two-thirds majority of the people.

THE IMPACT

Political leaders at most levels of city, county, and state government in California were less than enthusiastic supporters of Proposition 13. Many displayed their keen sense of business acumen by stating the evils of this piece of new expense control legislation. There were many heated debates on the issues, both pro and con. The statement was made that the issues under discussion were much to complex for the average person on the street to understand. There may have been some truth in that statement; however, it may have been equally hard for political officials.

One thing that could be understood was this well-known and proven fact. All monies had to be raised from a taxable base. Real estate provided a substantial part of the income potential for city and county governments. It was not uncommon for the appraisal rates and taxes due to be escalated ahead of inflation growth each year. The average taxpayer had no input into this process. One thing was certain—a larger bill for property tax could be expected each year. With the help of the U.S. Postal Service the bills were delivered, in most cases, on time. From that point on the money began to roll into the Tax Collector's office.

The introduction of Proposition 13 has reversed that trend! In summary, an idea had been developed, discussed, reviewed, and voted on by the people. The result, one of the largest cost reductions in the public sector had been adopted in a legal manner.

A Harbinger?

The Proposition 13 vote has been depicted as the harbinger of tax limitation proposals in other states. To a certain extent, that could be true. But there were five major factors that contributed to the California vote that were unlikely to combine to produce so dramatic an outcome elsewhere.

The first factor peculiar to California is that the state government there had accumulated a $5 billion surplus. That fact alone had angered some voters who thought the money was being hoarded in Sacramento, and it also encouraged voters to support the property tax limitation scheme on the grounds that the state surplus could help cushion the impact.

Second, the Proposition 13 vote was prompted at least in part by the fact that California had an unusually high property tax burden, which had been rising at an exceptionally rapid rate. In 1943, the property tax in California was below the national average, consuming about 3.3 percent of residents' income. In 1976, however, it had climbed to 6.4 percent of income, a level 42 percent above the national average.

Third, the state and local tax burden in California was exceptionally high. It was about 20 percent above the national average.

Fourth, California had one of the easiest systems for initiative proposals in the country. In most states it would be much more difficult to enact constitutional amendments limiting taxes.

Finally, residential property values had been climbing at very rapid rates in California to make the resulting climb in taxes even more oppressive.[1]

Congressional Reaction

The approval on June 6, 1978, by California voters of Proposition 13, a state constitutional amendment sharply reducing property taxes, rekindled a debate in Congress over federal spending and tax policies in 1978. "The mood of Washington was summed up in two words—Proposition 13," observed Senate Budget Committee Chairman Edmund S. Muskie, D-Maine, summarizing the way the California decision had become woven into the fabric of politics in Washington, denounced by some, and praised by many—although for very different reasons.

Few observers expected the California vote to lead to changes as dramatic in other states, or in Washington, as they were in California, where the voters slashed real estate taxes by almost 60 percent—about $7 billion—in one stroke.

That was partly because the Proposition 13 movement had some characteristics peculiar to California. But it was also because lawmakers in Washington did not fully agree on what the nation implications of the vote were. In general, Proposition 13 was considered likely to reinforce a trend toward some restraint in spending. But beyond that, there was considerable disagreement over the best way to address the frustrations expressed in the 2-to-1 vote in support of the tax limitation initiative.

Some viewed the vote as an attack on the size of government itself.

They insisted on substantial spending reductions, regardless of the effect on government services. But others, predicting a backlash in California when voters there realized the extent to which Proposition 13 would reduce services, argued that the real message of the vote was that government must find ways to provide services more efficiently so that taxpayers would get more for their tax dollar.

Congress was even more divided over the federal tax implications of the Proposition 13 vote. Some have depicted it as a demand for lower federal taxes. But others believed it was mainly a reaction to inflation, which they argued could be worsened by large tax cuts that would swell the federal deficit.

Moreover, some viewed the vote as part of a broad-based tax revolt, while others saw it mainly as a reaction to California's unusually high property taxes. Behind that difference in interpretation lay a sharp disagreement in Congress on which federal taxes, if any, should be reduced.

The reaction to Proposition 13 is in the eyes of the beholder. Many members of government at all levels—city, county, and federal—were able to read the message as a "preview of coming attractions." The people had spoken and were demanding to be heard. The June 1978 message was just one of a number of related statements that have been heard. Other related statements are being heard in Detroit, Boston, and Indianapolis.

Long-Term Trends

To the extent that Proposition 13 reflects a national reaction against rising government spending and taxes, there was some justification for it. Federal spending has climbed steadily from about 18.7 percent of gross national product (GNP) in 1958 to an estimated 22.6 percent in 1978. Federal taxes have climbed as well. While they only surpassed 20 percent of GNP twice since 1958—in 1959 and 1970—they have been rising rapidly since 1976. In 1976, they comprised 18.4 percent of GNP, climbing to 19.4 percent in 1977 and to an estimated 19.6 percent in 1978.

State and local spending and taxes also have jumped significantly. Spending rose from 8.2 percent of GNP in 1959 (excluding federal aid) to 11.5 percent in 1974, although it dropped to an estimated 10.6 percent in 1977. State and local taxes, in the meantime, have risen from

7.25 percent of GNP in 1960 to 9.66 percent in 1976. There were some signs that trend could be slowing. President Carter, for instance, pledged to bring federal expenditures down to about 21 percent of GNP.[2]

In summary, the taxpayer has been and is still looking for a businesslike approach in the operation of government at all levels. Perhaps we are closer to this operational mode than we have been in the past.

CHALLENGES IN SELECTED CITIES

Three Years Later

June 23, 1981, was another milestone in the era of deficit spending for the city of San Jose, California. After weeks of debate and meetings, the San Jose City Council approved a $479 million budget for 1981-82.

The budget relies on $4.4 million in one-time revenues to expand the police and fire departments, and to prevent wholesale elimination of recreation and library services. The total spending package includes $162 million for operations, $136 million for special-purpose facilities, and $181 million for capital improvements. The total budget is about 18 percent larger than last year's $407 million.

The operating budget, from which police, fire departments, parks, libraries, recreation, and other neighborhood services are funded, will increase just 1 percent over last year's $161 million.

The council, which had set public safety as its top priority, decided to add 25 police officers and eight firefighters during the coming year, giving the police department about $35.1 million compared to $34 million this year, and the fire department about $19.9 million compared to $19.6 million this year.

Parks and recreation will receive slightly less money this year compared to last—$11.1 million compared to $11.2 million—and libraries will receive $4.5 million compared to $4.8 last year. Refer to Table 12-1. Although the reductions are relatively small, they gain magnitude because of inflation and increased demand for services. In fact, the staffing level of the parks and recreation department will drop from 545 this year to 471 next year, and the library staff will fall from

Table 12-1. San Jose Budget Comparisons.[a]

DEPARTMENT	1980–81	1981–82	IMPACT
Police	$34.0 million 1159 employees	$35.1 million 1166 employees	Provides for 25 new officers on the street by May 1982.
Fire	$19.6 million 672 employees	$19.9 million 680 employees	Provides for eight new firefighters
Parks and Recreation	$11.2 million 545 employees	$11.1 million 471 employees	Reduced maintenance overall. Fees for classes at community centers will be raised from $2 to $11. User fees increased for baseball diamonds and other park activities.
Public Works	$ 8.1 million 241 employees	$ 9.3 million 242 employees	Utility excavation permits will be increased from $270 to $430. Rates for sewer trenches will also increase.
Neighborhood Maintenance	$ 8.1 million 247 employees	$ 9.2 million 239 employees	Wide-ranging fee increases. Tree trimming, formerly free, will cost $36. Sidewalk repair will cost $50 plus the cost of a contractor.
Library	$ 4.8 million 222 employees	$ 4.5 million 204 employees	Main library hours will be reduced from 60 to 40 hours a week.
Mayor and City Council	$910,000 28 employees	$910,000* 26 employees	*Does not include pending increases in council salaries. Ten percent cuts in individual nonsalary items. $50,000 formerly in this budget for a citywide audit was transferred to another account.
Neighborhood Preservation	$1.26 million 55 employees	$1.50 million 58 employees	Substantial fee increases for ordinary inspections and code enforcement. Multiple housing permits increased from $4.25 to $11 per unit. A doubling in prices for animal and fowl permits, kennel permits, fence variances. No change in mobile home inspection fees.
Planning	$1.09 million 46 employees	$1.1 million 43 employees	Suspension of development of comprehensive energy plan. Elimination of staff help for new downtown zoning district. For other fee increases, see private development.
Private Development	$3.6 million 130 employees	$3.6 million 130 employees	Huge fee increases, with final levels still to be determined. Proposed is a $1000 fee for general plan changes and substantial penalties for Williamson Act cancellations. Also under consideration is a near doubling of fees for developing private homes, from $676 to $1350 for a $100,000 home.

[a]*Source: San Jose News.*

222 to 204. The council approved increases in fees and charges in the city that are expected to raise $5.3 million in revenues. Most heavily hit will be the development industry and, in turn, buyers of new homes.[3] Scheduled increases include, for example, $1500 for a one-acre planned development zoning compared to $400 now; $2650 to cancel a Williamson Act land contract compared to no charge now, and as much as $2450 for the most expensive general plan amendment request (more than three acres and outside the urban service area or in a hillside area).

There will be higher charges for overdue library books during the coming year, as well as for park reservation and entry fees, tree trimming, building permits, and scores of other charges paid by citizens. There will be no charge for library cards, however, and homeowners who trim their own street trees will not have to purchase a city permit.

Supplies and materials were reduced through the city by $625,000. This cut was made over protests by the city manager. This reduction amounts to about $1.00 per head count of the city's 630,000 population count. Some observers would doubt this as a real cost-cutting effort.

Deficits Abound

Santa Clara County is no exception when it comes to continuing the trail of budget deficits. The 1981-1982 budget is anticipating a $32 million shortfall. The director of management and budget stated that no real progress was being made in dealing with the real problem, developing a realistic budget. One recent article cited a reduction of 55 county jobs for a savings of $7 million, only a slight dent in the $32 million deficit.[4]

Untapped Savings. Santa Clara County could save $633,000 annually if it handled its car pool more efficiently, according to a county-commissioned audit. The county has too many cars that are maintained at too high a cost, the audit, conducted by Harvey M. Rose Corp., concluded. Auditors found county mechanics take 39 percent longer to do their work than industry standards dictate. The report recommended that incentives, such as a four-day work week, be offered to increase county mechanics' productivity. An example of "questionable" practices found by the auditors was a vehicle that had

$963 worth of body work performed on it in December 1980. It was sold for $800 five months later. Auditors also criticized the county garage management's practice of contracting for outside repair work at a higher cost than would be paid if the work were done by county mechanics.

The auditors said the county has 186 more cars than it needs, partly because county departments have been allowed to request new cars whenever they wish. In 1974, for example, the county bought 135 Plymouth Valiants. Forty-two were discarded before the required mileage or age. "These low mileages were due primarily to user dissatisfaction with the vehicles' manual steering system and the garage's consequent reluctance to force use," according to the audit report. To compensate for this lower-than-expected utilization, additional vehicles were purchased in subsequent years.

Although other audit plans were not announced, they may in fact be worthwhile.[5]

Realignment Savings—Indianapolis

A reorganization of the Indianapolis Fire Department, reported to be a budget-cutting move, was announced in mid-May 1981. The Public Safety Director said the move calls for a realignment of the city's fire districts. In another move, a former arson chief was returned to headquarters to head up the arson division.

There will be a cut in the fire districts. Sources within the department say the districts will be cut from the present six to five or four districts. The move will mean a cut of three to six district fire chiefs, a savings to the department of $69,000-$138,000 a year. Each district chief presently receives a salary of $23,000. The savings does not include the salaries paid the three to six lieutenants who serve as the district chiefs' aids. They will be transferred to other duties within the department.

The objective of the changes is to bring about new cost savings—increased efficiency that translates into productivity. This approach will keep in 1982. The proposed realignment will prevent any fire fighters from losing their jobs if no future cuts are required. The reorganizations were the result of an investigation associated with the reported theft of property at fire department repair shops.

The city in 1981 was expecting about a $13 million decrease from the approximately $50 million in federal funds received in 1980. Sources close to the city's budget operation feel that Indianapolis can cope with the reduced revenue better than many other large cities. [6]

Detroit Versus Bankruptcy

A constant fight for survival has been under way in Detriot since the riots of 1967. Since that time city government has been learning to live with the loss of one-third of its population and associated revenues. Even with the auto industry decline, the city appeared to be holding its own. One visible symbol of this fact is the new Renaissance Center. The new symbol of glass and steel was constructed at a cost of $337 million.

The Republican Convention of 1980 was held at the new Renaissance Center in downtown Detroit. Many preconvention skeptics among the delegates and journalists left town with a feeling that a partial recovery was in progress.

By mid-1981, Detroit found itself looking bankruptcy in the eye for the second time in 50 years. City employees were being asked to accept pay cuts. Informed sources indicated that more layoffs may be required. The city's dream of a rebirth was turning into a nightmare and economic reality:

- The auto industry sank to new lows. Automakers lost $600 million in the first quarter of 1981, the largest losses ever up to that point.
- Unemployment soared to 13 percent, nearly double the national average. State labor analysts estimate that for every auto assembly job lost, five associated support workers are idled.
- City residents became very dependent on some form of government assistance. An estimated 60 percent of the city's 1.2 million residents were receiving unemployment benefits or some other form of aid.
- Business activity reflected the slowdown in the economy. Per capita retail spending was about 62 percent of the national average.
- City expenses grew about 41 percent in the previous three years, while revenues increased only 18 percent. The result: a deficit that continued to grow.

In mid-May 1981 the mayor outlined an action plan to keep the city from going broke:

- The city's income tax, already one of Michigan's highest, would be doubled to 4 percent. This would produce an estimated $94 million.
- Pay of city employees would be reduced in the range of 5–7 percent. Estimated savings would be in the range of $76 million.
- Local banks and employee pension funds would be asked to purchase $132 million in bonds to cover accumulated debt. This action would require approval of the state legislature.

At the unveiling of a $1.6 billion budget, Mayor Coleman Young had this to say: "There are no hats with rabbits in them, and no magician is waiting in the wings. We are fresh out of magic and miracles."[7]

The Boston Plight

In the midst of an unprecedented building boom in mid-1981, Boston is very close to the brink of a financial bust. While construction is under way on seven new hotels with 3400 rooms, four skyscrapers are also being built. Approximately 4 million square feet of new and renovated office space is being created. However, at the same time, the 64,000 public school children in the city attend school under a court order. Their plight is to be resolved by the city council. The issue at stake is how to keep operating as outgo of funds continue to exceed revenues.

People in general are confused. They hear conflicting views on the status of the future of Boston. One view would indicate the city is in great shape: a livable city of theaters and parks, a once-ramshackle maze of abandoned warehouses polished up and now fancy, expensive waterfront condominiums, refurbished and bustling Faneuil Hall Market drawing 1 million visitors a month. An opposing view might be that the city is falling apart: a city more than $75 million behind in property tax repayments ordered by a state court and looking ahead to a $118 million drop in revenues next year, with 5000 of its 12,000 municipal workers facing layoff.

And its increasing inability to borrow money via the bond market has meant that Boston's spending for bridges, tunnels, and roads is roughly half what it was early in the decade. As a result, three-quarters of the city's bridges are in need of major repair, and other repairs are way behind.

The mayor blames the financial problems on Proposition 2½, approved by Massachusetts voters, including nearly 90,000 Boston residents, in November 1980. The measure limits property tax to 2½ percent of market value. With a tax rate approaching 10 percent, the city was hit hard.

"Boston has a thriving economy caged in an antique governmental and government finance system," said the commissioner of city hospitals and head of a special budget committee advising the mayor.

"Since fiscal 1975, the city has spent $121 million above appropriation," said the associate director of the Boston Municipal Research Bureau, which monitors the city's financial health for the local business community. "The city has had the practice of spending beyond appropriation and then raising taxes to cover it," according to the associate director, "Last year in 1980 the city spent $25 million above its budget."

In the past, the city could roll over deficits to the next year and raise taxes. Proposition 2½ prohibited that mode of operation.

The school crisis is a prime example of the problem. The school system has overspent its budget every year since 1972; it has survived six violent years of court-ordered busing and a 25-percent drop in enrollment.

The city's problems are compounded because it is property poor. Only 20 percent of the greater Boston area is actually in Boston; other major cities take up about two-thirds of their metropolitan area.

In summary, Mayor Kevin White states the situation in the following manner. "People are wondering . . . one minute they hear Boston is in great shape and the next minute your hear she's falling apart."[8]

The San Leandro Experience

San Leandro, California, is a small east bay community that for years has been in the shadow of Oakland and San Francisco. In the period from 1970 to 1982 the population decreased from 74,400 to 64,640. This trend is not unique for cities both large and small. In order to focus on the problem, let us view a few critical items in 1970–1971 and the same ones in the 1981–1982 period.

FISCAL YEAR	OFFICIAL ASSESSED VALUE	TOTAL BUDGET	CAPITAL IMPROVEMENT IN BUDGET
1970–71	$262.9 million	$ 9.567	$1.233
1981–82	$500.1 million	$27.817	$2.322

Assessed value in the period increased about 90 percent, the total budget increased about 290 percent. How does a city cope with this problem?[9]

Coping with the Problem.   San Leandro, like most other cities, has been beset from many directions by forces beyond the control of the city council. The federal and state governments have created uncertainty regarding funding, continuity, and regulation of federally financed programs and projects. At the same time, the State has also continued to mandate, through laws passed by the legislature and administrative regulations, that the city perform additional tasks requiring additional staff time and expenditures, without provisions for full reimbursement. And finally, constitutional, voter-approved revenue and expenditure changes have had a dramatic impact on the city's continuing financial ability to perform.

The voter-approved tax limitation initiatives and related laws and regulations have had the most significant impact on the city's finances. Proposition 13, as anticipated, had both an immediate and a delayed impact. The immediate impact was a 12-percent reduction in San Leandro's total revenue (approximately $1,750,000). The city, in a carefully planned manner, immediately accepted this revenue loss and adjusted to it by reducing capital expenditures and some city programs, resulting in significant personnel reductions. Within a relatively short period of time the city adjusted itself to the new financial realities. It reduced expenditures and began preparing future budget strategies based on anticipated revenue losses. The delayed impact of this revenue loss is now being further felt by San Leandro and other local government agencies supported partially by property taxes. The property tax had previously been a significant and responsive source of revenue. With the property tax limited to 1 percent of value, and limited to a maximum 2 percent annual increase in assessed value, there was a built-in certainty that this source of revenue would fall steadily behind the inflationary increased costs of operating the city. High construction costs and interest rates together with a dramatic reduction of property resales have further reduced the potential growth in property tax revenue. Refer to Figure 12–1.

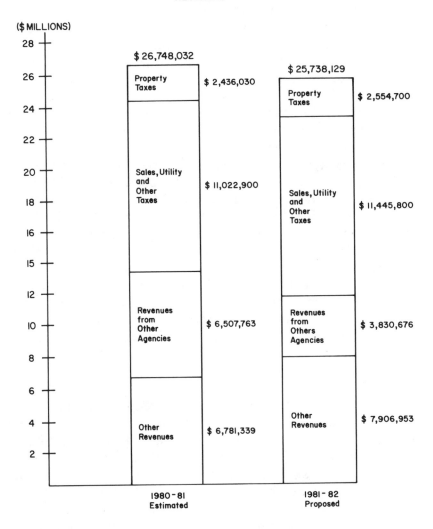

CITY OF SAN LEANDRO
TWO-YEAR BUDGET COMPARISON
REVENUES

Figure 12-1.   City of San Leandro: two-year budget comparison revenues. (*Source:* San Leandro budget.)

The city council now controls (i.e., can increase or decrease) only approximately one-third of the city's revenue. This one-third control can be exercised primarily for self-supporting activities, (e.g, golf course green fees, refuse collection rates, and sewer user fees). Thus, it does not have the ability to price its other services to meet the cost of providing those services. This situation is made more difficult by the rapidly inflating costs of what the city uses to provide the services. General fund revenues, in total, are only slightly responsive to the impact of inflation. As a result, the city continues to experience a rapidly widening gap between total revenues and expenditures.

The adverse impact of the factors affecting the city's long-term ability to finance its operation is further compounded by the uncertainty of state aid to local government. It should be noted that the state is facing a deficit situation for 1981–1982. Its alternatives will be to cut its own operating expenses (programs and personnel), cut aid to local government, or increase revenues. As noted, with Proposition 13 the city's property tax revenue was immediately reduced by $1,750,000. The state provided "bail-out" money of about one-half of that loss. Now the city is faced with the governor's 1981–1982 proposed budget, which would take away $946,000 from the city.

This is a significant departure from the past practice of recognizing that certain types of revenues, such as property tax, should be returned to local government for the provision of property-related services. This apparent change in state thinking creates great uncertainties as to both the source and amount of some major city revenues. At this time the clearest signal as to the future of state aid is the governor's budget.

Revenue Evaluation. Some city revenues are not responsive to inflation, and some revenues are only slightly responsive. In addition to the difficult inflationary aspects of the economy are its current "flatness," as well as the unknown impact of federal fiscal and monetary policies. The city cannot either control the forces that determine those revenues, or know where the economy is going. The sales tax is a good example of this. Over the past eight years it has fluctuated year to year between a 1-percent decrease to a 14-percent increase. Currently each 1-percent change in sales tax revenue can change general fund revenue up or down by about $89,000. Thus, realistic but cautious estimates were a necessarily prudent approach to developing a budget.

Substantially all of the financial uncertainties mentioned in the above discussion have been with the city in varying degrees since July 1, 1978. Refer to Figure 12–2 for a graphic summary of proposed

CITY OF SAN LEANDRO
GRAPHIC SUMMARY OF PROPOSED APPROPRIATIONS
1981-82

GENERAL CATAGORY	DOLLARS	PERCENT
General Services City Council, City Manager, City Clerk, City Attorney, Finance, Personnel Community Development, Community Relations, Insurance and Non-departmental Services	$3,821,572	13.7%
Public Safety Fire, Police, Communications, Disaster Preparedness and Animal Control	7,089,876	25.5%
Public Works Administration, Engineering, Code Compliance, Street, Tree, Park, Lighting, Signal & Sewer Maintenance and Parking	3,708,257	13.3%
Retirement	2,470,310	8.9%
Leisure - Recreation, Library & Marina	1,318,379	4.7%
Street Gas Tax (Sec. 2106, 2107 & County)	1,061,695	3.8%
Capital and Business Improvements	2,322,518	8.4%
Dept Service	139,858	0.5%
Enterprise Refuse, Golf Course & Water Pollution Control	4,053,079	14.6%
Federal & State Grants H.C.D.A., Revenue Sharing, Special State and Federal Grants	1,832,055	6.6%
PROPOSED CITY APPROPRIATIONS	$27,817,599	100.0%

Source: San Leandro Budget

Figure 12-2.   City of San Leandro: graphic summary of proposed appropriations, 1981-1982. (*Source:* San Leandro budget.)

appropriations for the 1981–1982 period. An examination of the proposed revenue and planned expenditures shows a deficit of $2.079 million, or approximately 8.1 percent.

Not a bad showing when compared with a cross section of cities around the country. Perhaps one key item has been overlooked: the relationship of the Consumer Price Index (CPI) and city revenues. An example of this is shown in Figure 12-3. In general, as mentioned before, cities will continue to be caught in a spiral of budgets that continue to rise. The key point is that revenues must be increased or expenditures reduced. A realistic approach may be to work in the area of increasing revenues while looking for new and innovative ways to reduce costs. [10]

Boise, Idaho—A New Technique?

One of the great strengths of any business—including government—is its ability to change. In this case change deals with the measurement of work done in an office environment.

It has been called insulting, degrading, and disruptive. But the man in charge says it is the best way he knows to weed out those who are not working. The new program is part of an efficiency move started by the Idaho Department of Health and Welfare office. Whistles are blown seven times each day in the office. Each time, 30 clerical employees are ordered to fill out forms describing what jobs they were performing at that moment.

The new chief of the welfare division is conducting a "random moment time study" of efficiency in his department. However, the secretaries and some other administrators are not happy with the arrangement. The study was implemented to aid in the selection of low-efficiency employees. The target in this case was three secretarial positions. The Welfare Department was looking for ways to reduce its budget $110,000; three employees would be a start.

Employees complained that the approach was nerve-racking. Some days the whistle would not be heard all morning. Between lunch and quitting time the seven blasts were very disruptive. In summary, a system such as this can be used, but it tends to create friction between the workers and management. [11]

Winston-Salem, North Carolina

Industrial engineering was introduced into the city of Winston-Salem in the early 1960s by the city manager. Initial efforts dealt with cost analysis, methods improvement, and work measurement. A point

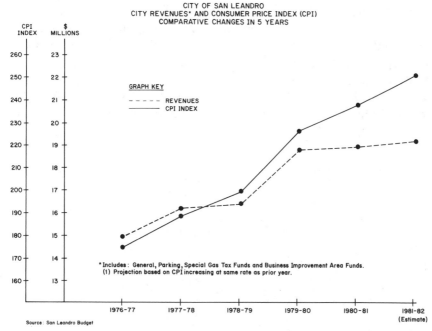

Figure 12-3.   City of San Leandro: City revenues and consumer price index (CPI) comparative changes in five years. Dashed line represents revenues; solid line represents CPI. City revenues include general, parking, special gas tax funds, and business improvement area funds. 1981-82 CPI projection in based on CPI increasing at same rate as prior year. (*Source:* San Leandro budget.)

was reached where data handling became a considerable problem. The focus of the industrial engineering effort shifted to updating the data processing facilities. Eventually, industrial engineering and data processing were combined into one department named Management Services. This new department concentrated on management information systems until 1977 when a new industrial engineering position was authorized by the board of aldermen. Currently, Management Services is composed of industrial engineering, systems development, and data processing operations.

Industrial Engineering Studies.   Since industrial engineering was reinitiated, the emphasis has been in the Public Works Department. Industrial engineering studies have been focused on maintenance activities in Streets and Water and Sewer Construction Divisions of Public Works. Generally, the goals have been to aid management with planning and controlling the individual work activities. Studies have dealt with establishing the optimum crew size and forecasting the workload for the particular activity. The studies follow a basic five-step approach consisting of defining a problem, collecting data, analyzing data, implementing solution, and evaluating results. Refer to Figure 12-4. The philosophy has been that improvement is the responsibility that is shared by operating departments and industrial engineering. In keeping with that, the supervision and management of

Define Problem

General Survey and Review
Select Area for Analysis
Define Scope and Objectives
Review with Department/Division Heads

Obtain Information

Announce to All Participants
Define Present Method
Select Work Unit
Collect Data

Analyze Data

Flow Charts
Statistical Analysis
Methods Engineering
Value Analysis
Systems Analysis
Cost Analysis
Engineering Economics
Operations Analysis

Implement

Informal Presentation
Prepare Report
Formal Presentation
Assist in Implementation

Monitor

Establish Control Reports
Follow-Up Observations
Document Actual Results

Figure 12-4.   Industrial engineering studies in local government. (*Source:* S. H. Owen, Jr., P.E., Management Services, Winston-Salem, North Carolina.)

the Public Works Department has been involved in each step of each study.

The studies have pointed out a general weakness in carrying out the planning and control responsibilities of management. In most cases, it has been possible to either reduce the crew size or make some other change resulting in a significant improvement in the unit cost of the service provided. The benefits of these individual studies have been significant, but the piecemeal approach to observing the various activities has been a weakness for industrial engineering.

Program Development. A decision was made to concentrate industrial engineering studies in one area of the Public Works Department. The decision was made to implement a maintenance management system in the Street Maintenance Division with a major goal of enhancing the planning and control capabilities. The system uses work standards to support the processes of work programming, crew scheduling, and performance reporting. The main efforts of the industrial engineer are to establish the standards and train the supervisors in programming, scheduling, and reporting.

The maintenance standard is established following the same five-step procedure as used in the industrial engineering studies. The standard defines why, where, when, how and by whom the work is to be done. The standard also prescribes the units the work is to be measured in, such as square yards, or linear feet, and an estimated amount of time for completing each unit of work.

Work programming and crew scheduling are the planning aspects of maintenance management. The work program is an estimate of work needs for a specific period of time, usually a year, expressed in terms of work units to be completed and man-hours required for completion. The work program serves two purposes. First, since the work program is a statement of manpower requirements to do a given amount of work, it is a valuable tool for preparing the annual budget. The work program then serves as an agreement between the chief executive officer and operating manager as to how much work is to be done and the staffing requirements to do that work. The second purpose of the work program is to provide a basis for crew scheduling which is a lower level of planning. [12]

PRODUCTIVITY A MUST

Up to this point a number of related problems that are common to all cities have been discussed. How to increase productivity is the challenge for all members of the public sector. This concept becomes even more important as revenues continue to drop and as the cost of services continue to rise. Productivity can truly be the key hope for city problems.

The following article by Richard L. Shell, P.E., and Dean S. Shupe, P.E., provides an insight into problem solving and improved results. A wide range of productivity projects are illustrated.

Galloping costs and plodding incomes afflict governments as well as private institutions. Similarly, a major hope they have to cope with the squeeze is to improve productivity

Richard L. Shell, P.E., and Dean S. Shupe, P.E.
University of Cincinnati

PRODUCTIVITY: HOPE FOR CITY WOES*

Service cut-backs, layoffs, reduction in capital improvements and possible financial default—these headlines evidence the growing plight of many cities across the nation. The typical municipality is being hard pressed to maintain its service-oriented, labor intensive functions in the face of continuing inflation while on a relatively fixed income base, Reference 1. "The problem is this: Over the past two decades, state and local spending, now running at $221.5 billion, has grown faster than any other sector of the economy. State and local expenditures exclusive of federal aid rose from 7.4 percent of gross national product in 1954 to 11.6 percent last year"(2).

According to a recent U.S. Census Bureau report, there were 2,506,000 municipal employees in October 1975, up 0.6 percent from 2,491,000 in October 1974. As of October 1975, cities were paying their employees wages at a monthly rate of $2.1 billion, a 7.3 percent increase from Octboer 1974. The Bureau also reported that municipal employment more than doubled since

---

*Reprinted with permission from *Productivity: A Series from Industrial Engineering.* Copyright ® American Institute of Industrial Engineers, Inc., 25 Technology Park/Atlanta, Norcross, Georgia 30092.

1946 when there were 1.2 million city workers. City payrolls in 1975 were nearly 10 times those of 1946 (3).

The revenues of local government have not kept pace with the spiraling expenditures. The bottom line result to date has been deeper budget deficits. Figure 1 illustrates the worsening financial condition since 1973. To relieve this mounting deficit pressure, most cities must either cut services or improve productivity.

The primary concern of municipal government should be to fulfill the collective needs and desires of its citizens. Productivity improvements permit higher quality or additional service at the same cost, or a uniform service at a reduced cost. In either case, productivity growth in municipal government largely depends on employees. As William Donaldson, City Manager of Cincinnati, recently put it: "I'm here to manage the city—to get the most out of our people and our dollars—and that's what I'm trying to do with the help of very capable people" (4).

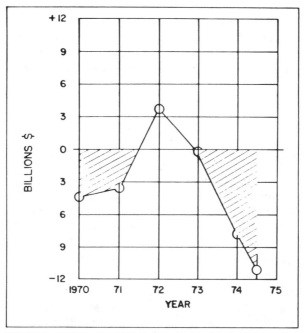

FIGURE 1. ANNUAL BUDGET SURPLUS OR DEFICIT FOR STATE AND LOCAL GOVERNMENTS.

Source U S Commerce Dept

Figure 12A-1. Annual budget surplus or deficit for state and local governments. (*Source:* U.S. Commerce Department.)

Areas for Improvement

Typically, the medium and larger municipalities provide a multitude of diverse citizen services. They include the major activities outlined below.

General Administration: Management and finance; Data processing and information systems; Purchasing and inventory; Personnel and labor relations.

Public Safety: Police services; Fire protection; Courts and criminal justice

Public Works: Solid waste collection and disposal; Streets and highways; Buildings and equipment maintenance; Sewer and water service; Inspections and energy conservation

Parks/Recreation/Libraries: Operations and maintenance; Library services

Social Services: Unemployment; Welfare; Volunteer programs; Public health

The Citizen services provided and the organizational structure for delivery of those services are functions of city size and other characteristics, e.g., socio-economic, demographic, and geographic. Given the wide range of municipal service activities, our cities offer manifold opportunities to pursue and implement productivity improvements.

Steps for Productivity

Characteristically, municipal management has considerable experience and is willing to accomplish productivity improvements. However, too often these managers lack the necessary technical tools to successfully analyze and solve the problems that deter productivity growth. The industrial engineering approach that has proven successful in the manufacturing sector should be more fully utilized in municipal government activities. The industrial engineer offers unique capabilities in such areas as: Work measurement and incentives; Engineering economy and capital budgeting; Data processing and information systems; Behavioral science and organizational development; Operations research.

The proper combination of technical, political, and behavioral skills are required for successful development and implementation of productivity improvements. Projects must involve the cooperative efforts of municipal management, the workers' bargaining representatives, and the workers themselves. When individuals in any organization, private or public, contribute to the design and implementation of desirable changes, that organization will realize greater productivity.

Barriers to Improvement

Major barriers to productivity growth in municipal government are twofold: technical and political. In most cases the industrial engineering approach can solve the technical problems. However, the political barriers existing in municipal government are more complex than typically found in the private sector. These barriers not only affect costs but also restrain freedom of action in developing, implementing, and maintaining productivity programs. In some cities, the political environment may be so difficult that productivity is actually retarded.

Political barriers may be broadly categorized into three groups: multiplicity of interested parties, characteristics of management, and legislative constraints.

Multiplicity of Parties: Many parties have interests in and impact on the operation of a municipality. These include: Citizens—individuals, organized special interest groups (e.g., Chamber of Commerce, neighborhood and civic organizations). City management—mayor, commissioners, city manager, department heads, and subordinate administrative personnel. Union/workers—individual employees, different bargaining groups, and workers excluded from productivity programs.

Characteristics of Management: In contrast to a private corporation, municipal management is often diffuse in its decision-making activities and transitory in leadership responsibility. There is no single overall authority in the typical city. Major decisions are usually influenced by at least several of the parties listed. A city manager serves at the discretion of a majority of the commission. Commission members are elected by the citizens and are influenced by public opinion as expressed individually or through organized special interest groups. Frequently the unions are politically active and exert strong influence on public opinion and election outcomes. Meanwhile, city department heads while organizationally reporting to the city manager, usually operate with a sense of independence and security provided through the civil service system.

Because of the nature of the appointive and elective process, city managers and commissioners generally experience a high rate of turnover, resulting in short-term accountability. Consequently, much time is expended in the "learning" process, with efficiency reduced, and long-range planning impaired.

Legislative Constraints: The successful planning, implementation, and maintenance of a municipal productivity program is also constrained by numerous legislative restrictions. Examples include: Civil service regulations governing personnel policies, e.g., hiring and layoff, job classifications and bidding, and wage levels. State laws restricting the right to strike or regulating bargaining, e.g., Ohio Ferguson Act and the New York Taylor Law. Laws specifying length of work day or denying payment of incentive bonuses.

What to Do

If the American city is to survive and provide a desirable quality of life for all its citizens, continued productivity improvements must be realized. Productivity programs should incorporate the following:

- Use the industrial engineering approach to realize benefits from new technology, and develop quantified alternatives for municipal government decision making.
- Solicit participation of the citizens and the workers in the planning, implementation, and maintenance of productivity projects.
- Share the resulting benefits of productivity improvement with all concerned; the citizens, workers/union, and management.

City Productivity Projects

Examples of municipal projects that have been reported by the National Center for Productivity and Quality of Working Life (5) or by the authors (1), include the following:

Police Services: As a result of sophisticated analysis of crime data, several cities schedule patrol units according to each sector's crime rate and accidents for that particular time period. These cities include Glenview, IL, and Chesapeake, VA.

Fire Protection: Several cities have worked with computer software (developed by Public Technology, Inc.) which determines how many fire station sites are necessary, where they should be located, the likely average response time, and the estimated number of alarms per city. These include Denver, CO, and Norfolk, VA.

Water Meter Reading: Fayetteville, NC, has operated a system of water meter reading via two-way radio communication with a central information office. Meter reading can be carried out in inclement weather, since field agents do not carry records. One dispatcher handles several field agents.

Parks: New York has implemented a city-wide parks maintenance improvement program based on work standards and performance reporting.

Waste Collection: Covington, KY, recently implemented new waste collection routes developed utilizing work measurement techniques. The new routes introduced time incentives and resulted in annual savings of $225,000

Table 1. Local Government Usage of Employee Incentives: A Summary of Survey Results from 509 Jurisdictions as of August–December, 1973[a]

	CITIES 25,000–50,000 (40 RESPONSES)		CITIES LARGER THAN 50,000 (315 RESPONSES)		COUNTIES LARGER THAN 100,000 (154 RESPONSES)		TOTAL OF ALL CITIES AND COUNTIES (COL. 1 + COL. 3 + COL. 5)		EVALUATION OF THE INCENTIVE PROGRAMS	
	1	2	3	4	5	6	7	8	9	10
	No. USING	% OF RESPONSES	No. USING	% OF RESPONSES	No. USING	% OF RESPONSES	No. USING	% OF RESPONSES	No. USING	% OF TOTAL PROGRAMS (COL. 9 + COL. 7)
Educational incentives	22	55	218	69	63	41	303	60	14	5
Output-oriented merit increases	17	43	135	43	61	40	213	42	22	10
Task systems	17	43	131	42	9	6	157	31	17	11
Suggestion awards	6	15	93	30	29	19	128	25	8	6
Attendance incentives	7	18	85	27	26	17	118	23	12	10
Variations in working hours	6	15	77	24	33	21	116	23	19	16
Safety incentives	4	10	73	23	14	9	91	18	5	5
Job Enlargement	2	5	54	17	17	11	73	14	4	5
Work Standards	2	5	37	12	27	18	66	13	0	0
Performance targets	4	10	41	13	10	7	55	11	0	0
Performance bonuses	0	0	27	9	5	3	32	6	4	13
Productivity bargaining	2	5	20	6	5	3	27	5	2	7
Competition & contests	1	3	14	4	0	0	15	3	1	7
Shared savings	0	0	3	1	1	1	4	1	2	50
Piecework	0	0	3	1	0	0	3	1	1	33
Others	0	0	23	7	7	5	30	6	0	0
None	7	18	30	10	47	31	84	17	—	—
Total Items Reported	90	—	1,034	—	307	—	1,431	—	111	8

[a] Source: National Commission on Productivity and Quality of Working Life.

representing 38 percent of original budget. The system provided the basis for a direct monetary wage incentive program.

Inspections: Several cities have trained inspectors to handle all buildings and construction permits. Consequently, productivity has been increased and motor vehicles are used less, resulting in lower maintenance costs and gas consumption. Cities include Scottsdale and Phoenix, AZ, and Dallas, TX.

Food Stamp Program: San Diego, CA, has analyzed the county's food stamp program and recommended a number of improvements which have been implemented. These include a new scheduling system which permits management to control the work load and a new prescreening process intended to eliminate ineligible clients prior to their being processed through the entire system. By eliminating unnecessary functions, as well as shifting and reordering work processes and redesigning forms, systems productivity (the number of clients the system can handle per day) has increased by 137 percent, error rates have decreased by 24 percent, and annual cost savings of $227,000 have been projected.

Management by Objectives: MBO appears to be gaining increased use by local government managers because of its emphasis on establishing objectives and evaluating subsequent performance in terms of tangible results. The following cities report some success with MBO: Cincinnati, OH, Ft. Worth, TX, San Diego, CA, and Little Rock, AR.

Cash Management: Baltimore County, MD, took competitive bids for county banking services and reduced the number of its accounts from 51 to 10, with a significant reduction in the average daily balance maintained. The resulting increase in investment earnings is expected to be between $50,000 and $100,000 (depending on interest rates).

Inventory Control: Oak Ridge, TN, has established a computerized inventory accounting system to maintain control and supply accounting information on 4,500 stock items and other materials purchased directly by users. The system generates output information for inventory control and develops information on materials cost for the maintenance accounting systems.

Citizen Input: Hamilton, OH, has purchased a bus for the Bring Ideas to City Hall (BITCH) program. The bus tours the City, requesting citizen suggestions, preferences, and comments on possible uses of Federal community development funds. The cost of the project is estimated at less than $2,000. The result has been increased government credibility and responsiveness, as well as increased citizen input in the government process.

References

1. Shell, R.L., and Shupe, D.S. "Improving Productivity of Solid Waste Collection Through Wage Incentives," *Proceedings,* Twenty-Seventh Annual Conference, AIIE, 1976.

2. "Borrowing Too Much to Keep Running," *Business Week,* Sept. 22, 1975, page 84.
3. "Municipal Jobs' Growth at 13-Year Low in '75," *The Wall Street Journal,* August 1, 1976.
4. "Pushing for Productivity," *The Cincinnati Post,* May 10, 1976.
5. *Guide to Productivity Improvement Projects,* Third Edition, National Center for Productivity and Quality of Working Life, Government Printing Office, July 1976.

REFERENCES

1. Simpson, R.P., "Taxpayers News," *United Taxpayers,* July 1978.
2. *San Jose News,* "Election 1978," June 8, 1978.
3. Trounstine, P.J., "San Jose Council OK's $479 Million Budget," *San Jose News,* June 24, 1981.
4. Yoachum, S., "$7.0 Million Cut From Budget," *San Jose News,* June 24, 1981.
5. *San Jose Mercury,* "County Car Pool Savings," August 20, 1981.
6. Birchfield, R., "City Fire Districts Reduced," *Indianapolis News,* May 14, 1981.
7. Gustavsen, J., "Detroit—On Bankruptcy's Brink," *Indianapolis News,* May 14, 1981.
8. Bayless, F., "Boston in Great Shape Also Busted," *Indianapolis News,* May 14, 1981.
9. City of San Leandro, "Proposed 1981-82 Budget," Office of Budget Administration, May 22, 1981.
10. Riordan, L.E., "Proposed Budget Highlights 1981-1982," City Manager's Office, May 22, 1981.
11. *San Jose News,* "They Get the Whistle While They Work," January 29, 1981.
12. Owen, S.H., "Industrial Engineering in the City of Winston-Salem," Productivity Seminar, October 14, 1980.

13
Conducting Cost Reduction Cases

Cost reduction is a team effort that involves the company, the employee, and the customer. In broad terms, cost reduction deals with a reduction of labor, material, floor space, energy usage, and expense costs.

Why do you need a cost reduction program? In a time of declining productivity nationwide, today's engineers and managers face constant challenges in their effort to remain competitive. These challenges include: the heavy flow of foreign goods into the United States, the threat of economic stagnation, and the resultant heavy trade deficits.

Companies, both large and small, have found that an effective cost reduction program can be the most effective tool in dealing with productivity in its downward trend.

How can a cost reduction program benefit you? An effective cost reduction program has three major areas of benefit: company benefits (improved productivity and quality and reduced production costs); employee benefits (personal recognition, job satisfaction, consideration for promotion); and customer benefits where the total impact of cost reduction is evident.

This chapter will present useful fundamentals and case studies of successful cost reduction programs that can be applied to a wide range of production operations, warehouse operations, and expense-related items. It will show you why, where, and how cost reduction ideas can be initiated, developed, and processed to a successful conclusion.

THE IMPACT OF COST REDUCTION

What is the impact of cost reduction? The answer to this question can be found in almost any industrial publication. At least one article in each will be devoted to cost reduction or productivity. This emphasis on cost reduction is usually brought about by the problem of meeting the demands of three groups:

- The buyer demands quality, service, and a competitive price.
- The worker requires wages commensurate with the change in the cost of living.
- The stockholder expects a fair return on investment.

In facing these problems, progressive companies must choose between lower profits or a reduction in manufacturing costs. In granting an increase in wages, the company must accept a decline in earnings or an increase in the selling price, or reduce the cost of manufacturing the product. Regardless of the end result, quality, service, and competitive price are the three factors that are still required regardless of the manufacturer's decision. Therefore, it is necessary to search constantly for cost reduction opportunities to meet these demands.

In seeking opportunities for cost reduction, consideration must be given to the division of responsibility for costs of customers' orders, land, buildings, machinery, materials, and labor. Our efficiency, and consequently our effectiveness, is dependent upon our ability to bring these elements together with the least possible time utilization and the greatest possible customer satisfaction.

Consequently the cost involved in each of the manufacturing elements must be a joint responsibility of labor, engineering, and management. One could go through each department and each function and show that an effective cost reduction program involves a team effort with everyone working toward a common goal.

COST REDUCTION CONSIDERATIONS

As a starting point look carefully into these ten items:

1. The use of less expensive materials, a reduction in quantities of materials normally consumed, or a reduction in the normal quantities of repaired, scrapped, or junked materials.
2. Improved, more efficient equipment to replace equipment worn beyond economical repair when the savings justify the excess expenditures, if any, for engineering and plant over those involved in a straight replacement.

3. Improved manufacturing processes and equipment including those involving changes or standardization in product design.
4. Economics in inventory, material handling, packaging, expense supplies, plant services, and maintenance for which engineering is functionally responsible.
5. Improvements effected in rearrangement and consolidation of shop organizations including reduction in floor space and/or reduction in material handling costs.
6. Improvements made effective at the time of a reanalysis of manufacturing methods and equipment to take care of increased requirements.
7. Manufacture of products or materials previously purchased or purchase of products or materials previously manufactured.
8. Computer applications controlling a productive operation.
9. Transfer of production work from one manufacturing location to another in order to reduce overall costs.
10. Investigations to reduce maintenance, expense items, power consumption, floor space, and telephone service.

Study this list carefully. It may be helpful to arrange the list in order from those areas that you are most familiar with to areas that may provide a challenge at a later date. Review the following four application examples. Look for the idea, the application, and the potential savings.

Moving Forward

The key ingredients of any successful cost reduction will be people, the work process, and automation.

1. People—Training, good supervision, and increased employee involvement in problem solving and communications. Actual ideas for improving can frequently come from lower levels of the organization doing the work.
2. Work process—With employee participation, careful examination of how you do work, and redesign of the process to improve efficiency. Understanding the work process better and applying people to it is the key.

COST
REDUCTION
POTENTIAL

APPLICATION:

LABOR . . . . . ●

MATERIALS . . ●

CAPITAL . . . . ●

ENERGY . . . . ○

EXPENSE . . . ○

IDEA: Evaluate make versus buy decisions on selected products and piece parts.

APPLICATION: A business machine company in the Bay Area purchased piece parts and assembled a key-slide assembly in thier own shop at a total cost of $13.75 each. No major problems were noted in the present operation. Rising labor costs were becoming a concern.

SAVINGS: By turning the total responsibility for the assembly to Ran-Rob, the final cost of the assembly was cut to $5.52, a savings of $8.23 per assembly. Yearly projections for 6,300 units provided savings of approximately $51,850.

SOURCE: RAN-ROB, Inc., Oakland, CA

COST
REDUCTION
POTENTIAL

APPLICATION:

LABOR . . . . . ●

MATERIALS . . ○

CAPITAL . . . . ○

ENERGY . . . . ○

EXPENSE . . . ●

IDEA: Evaluate the use of a robot for operations such as packaging, tending metal-working machines.

APPLICATION: General-use robots priced under $10,000 each can be utilized in a variety of material-handling and machine-loading applications. In general, a robot is capable of a continuous body sweep of 90° per second. Shoulder sweeps of 180° and elbow sweeps of 210° are possible. Each axis has its own air-servo motor.

SAVINGS: A robot can pack material as fast as the average human packer without boredom or fatigue. Minimal downtime and consistent performance make the robot ideal for some applications. Labor savings can range from $5,000/$20,000 per year depending on the industry segment.

SOURCE: Mobot Corporation, San Diego, CA

COST
REDUCTION
POTENTIAL

APPLICATION:

LABOR . . . . . ●

MATERIALS . . ●

CAPITAL . . . . ○

ENERGY . . . . ○

EXPENSE . . . ●

IDEA: Investigate foam in place-packaging techniques for shipping fragile items.

APPLICATION: A Canadian manufacturer of decorative ceramic lamps and figurines had been using shredded paper in the packaging operation. Three workers were required to pack one day's production - 150 lamps. Damage claims were exceeding $3,000 monthly. The use of Instapak foam-in-place reversed that trend.

SAVINGS: The packing crew has been reducted from three to two. Two 55-gallon drums have reduced the need to handle and store 10 tons of shredded paper. Yearly savings in the area of $60,000 have been obtained due to decreases in labor, breakage, and packing material.

SOURCE: Sealed Air Corporation, Saddle Brook, NJ

8

COST
REDUCTION
POTENTIAL

APPLICATION:

LABOR . . . . . O

MATERIALS . . O

CAPITAL . . . . O

ENERGY . . . . ●

EXPENSE . . . ●

IDEA: Evaluate ways to convert the disposal of waste such as wood, paper, cartons, cloth, rags, and hospital waste into a cost reduction.

APPLICATION: 1) A clothing company generates approximately five tons of waste daily. The waste is incinerated and converted into the equivalent of 150,000 gallons of fuel oil per year. 2) A large hospital generates the equivalent of 140,000 gallons of oil per year from waste material. The conversion ratio, waste to energy conversion, is about one ton of waste to 100 gallons of fuel oil.

SAVINGS: The conversion of waste material provides two avenues for cost reduction: 1) The reduced need for hauling the waste to a landfill site . . . The two examples above are in the $30,000 to $50,000 range. The energy generated can be the equivalent savings of $.86 per gallon of reduced fuel oil consumption.

SOURCE: Kelly Company, Milwaukee, WI

3. Automation—Tools: computer-aided drafting and design; word processing equipment; data base management systems. Automation must be approached in terms of support rather than replacement. Automation can improve the quality as well as the volume of useful work people perform.

A CASE IN POINT—MATERIAL USAGE

To offset part of the rising cost of tantalum, North Carolina Works engineers are working to reduce the amount used in capacitors. They have succeeded to the extent of saving $2.6 million annually.

Among the tantalum-conservation steps made during 1980–1981 on tantalum capacitors are these:

- Change to smaller-diameter tantalum wire for the positive terminal.
- Reduction in the amount of tantalum powder in the sintered positive electrode, made possible by process changes.
- Use of higher-capacitance tantalum powders, permitting a 15-percent decrease in tantalum powder consumption on many capacitor models.

Prices for raw tantalum metal—used in tantalum electrolytic capacitors manufactured at the North Carolina Works—have increased from about $70 per pound in early 1979 to $225 per pound in 1981. Tantalum, which is used in machine tools, aerospace structures, and medical products, is a semirare metal found outside the United States—in the Far East, Africa, South America, Australia, and Canada.[1]

The Approach

How to start in the pursuit of a cost reduction case is usually the hardest part of the process. Some engineers favor a checklist that can be reviewed—this seems to start the creative juices flowing. Others use different approaches such as holding the product in their hand or placing the item on their desk as a constant reminder. Develop your own technique for generating ideas. The following checklist of ideas may be helpful.

CHECKLIST OF COST REDUCTION IDEAS

The engineer and production supervisor both come in close contact with operating methods and conditions and are in a favorable position to be able to recognize opportunities for improvement. Let us review a brief checklist of cost reduction ideas.

I. *DESIGN*
   A. Eliminate parts or finishes.
      1. Doubtful function or duplication of function exists.
- Reduce number of screws.
- Reduce number of rivets.
- Eliminate washers or insulators.
- Eliminate leads by rewiring.
- Eliminate components.
- Eliminate brackets.
- Finish not required.

   B. Combine functions.
      1. Change design of part to perform a function formerly requiring several parts.
   C. Change physical shape of parts.
      1. Reduce size.
      2. Reduce thickness.
      3. Reduce scrap or excess material.
      4. Reduce operations by changing shape.
   D. Review tolerances and design requirements consistent with functions.
   E. Substitute materials or finishes.
      1. Aluminum for brass, or plastic for aluminum.
      2. Plated steel for other steels.
      3. Machinable steels for less machinable steels.
      4. Brass for nickel silver.
      5. Lower-grade critical materials for higher grades.
      6. Powdered metals for machined metals.
      7. Plastics for metal.
      8. Die castings for machined parts.
      9. Metallized materials for fabrics.
     10. Zinc for nickel.
     11. Enamel for plating.
   F. Use commercial parts or apparatus.
      1. Substitute standard commercial parts or apparatus for your own design.

G. Substitute high-production low-cost parts for low-production high-cost parts.
   1. Screws, rivets, eyelets, terminals, etc.
H. Redesign to utilize improved fabrication processes.
   1. Impact extrusions or injection molding.
   2. Epoxy resin castings for plastic parts.
   3. Printed circuits for complex wire and soldered circuits.
   4. Adhesive fastenings.
   5. Ultrasonic or cold welding.
   6. Machine assembly versus manual assembly.

II. *FABRICATION*
A. Eliminate unnecessary operations.
   1. Deburring, redrilling, polishing, reaming, adjusting, detailing, etc.
B. Combine operations.
   1. Progressive punch and dies.
   2. Dial feed machines.
   3. Special attachments for screw machines and punch presses.
   4. Conveyor operations.
   5. Special multiple operations machines.
   6. Multiple parts tool.
C. Substitute facilities.
   1. Machine operations for hand operations.
   2. Power screwdrivers for other types.
   3. Automatic detail or inspection machines.
   4. Higher-production machines—screw machines and presses.
   5. Automatic feeds for second operations.
   6. Separation of scrap from parts at operation.
   7. Automatic cleaning at operation.

III. *MISCELLANEOUS*
A. Reduce merchandise losses and repairs.
   1. Reduce repairs and dropouts by changing process.
      • Completely cover instructions and procedures for complex setups in layout. (Use sketches if necessary to illustrate.)
      • Investigate capability of facilities and components.
B. Review and improve repair procedures.
      • Provide adequate tools and machines for repair operations.
      • Improve repair sequences in complex assemblies.
   1. Materials handling.
      • Conveyorizing—transfer conveyors between machine and operations, etc.

- Bulk handling—tank storage and pneumatic transfer of materials, etc.
- Improved material flow.
2. Packaging.
   - Cheaper packing materials.
   - Automatic packaging—evaluate contour foam in place.
   - Automatic stamping and labeling.
   - Bulk packaging or palletizing.
   - Reusable packing material.
3. Inspection.
   - Reduce inspection by use of a sampling plan.
   - Review sampling plans and process averages with view to reducing inspection effort.
   - Combine operations in one gauge to reduce inspection effort.
   - Use indicator-type gauge for better control of the process.
4. Work simplification.
   - Add radii, chamfers, bevels, tapers, pins, plugs to reduce positioning and assembly time.
   - Arrange work for shortest transport distances.
   - Specify operations so that both hands can be utilized simultaneously.
   - Use graphic charts to show the old versus new method.

FACTORS REQUIRED TO DETERMINE COST REDUCTION SAVINGS

Cost reduction cases include savings of direct labor and material as a result of changes in method, process, type of material, or floor space. The changes must be approved and implemented before the savings can be credited and the case closed.

Labor Rates

Labor rates used for calculating savings are based on those in use at the time the saving goes into effect and include standard labor increment factors as published by the engineering and accounting departments. Where a saving is made on an operation covered by wage incentive rates, the labor portion of the saving must be reflected by a change in the rates. Credit hours should be used in calculating savings.

Savings resulting from installing wage incentives, or from covering a day work operation with a wage incentive rate, or the result of motion improvements during the process of rate setting are not credited as cost reductions.

Savings on day work operations that remain on day work must be definitely measurable as a reduction in the operating cost. Merely making a day worker's job easier is not necessarily a saving.

Where a cost reduction results in an increase in "expense" labor, the increased cost must be subtracted from the calculated savings.

Volume

Estimated savings are calculated on the volume of the involved product made in the current 12 months.

Material

The material element includes raw material, fabricated piece parts, and component apparatus received by the shop for further processing. The cost of material received from an outside supplier is the purchase price. The cost of piece parts or component apparatus received for further processing is calculated by adding the raw-material cost of the item to the unloaded labor cost already expended on the item.

Floor Space

Approved savings on floor space in buildings may be calculated at dollar value per square foot available. The space must be vacated and reassigned for another purpose.

Expanded Potential

Up to this point we have looked at cost reduction application in the production and warehouse environment. These two areas have been valuable sources of cost reduction over the years. The shift in work-force makeup is away from the traditional production- and ware-house-related jobs. Office jobs of all types are on the rise. White-collar jobs will continue to develop at a faster pace than traditional blue-collar jobs.

Recently I talked with an engineering manager of a computer company in the Bay Area. He told me that there were a total of 2100 people in the worldwide operation of his company. Of that total, only 700 were engaged in production and warehouse operations. The remaining 1400 personnel are classified as indirect or expense employees. In terms of cost reduction potential this segment of the work force offers a yet untapped source of savings. Take a look at your company, your organization, your department. What cost reduction potential do you think is possible?

Strange as it may seem, the same cost reduction guidelines will apply. The same considerations for expense control apply. These guidelines can be applied to all segments of business and service operations. Why is this not currently being done? The main reason for lack of progress in this area is inability to get started. In order to get started, management must be willing to state that the status quo is not good enough anymore. They must say to employees—something that management does not say often enough—we need your help to improve the operation.

This request for help can be verbal or written. The written method is usually more effective. Keep in mind that the end objective is to involve people, people who want to share their ideas. To improve the quality of the input ideas tell the people what's in it for them.

APPLICATION AT CENTRAL METHODIST COLLEGE

Let us review a program of this type that was started at a small college in the Midwest. This program is universal in nature. Dr. J. A. Howell is President of Central Methodist College in Fayette, Missouri. Fayette is midway between Kansas City and St. Louis. Central College is the result of unifying educational ventures carried out by the Methodists. Prior to 1841 the Methodists operated Howard High School, an outstanding private coeducational school. In May 1961 the name of Central College was changed to Central Methodist College.

Central Methodist College is a gift-supported institution, deriving no income from taxes or other public funds. Tuition and fees cover only part of the instructional cost each year. The remainder of the operating budget is derived from Endowment Fund income, alumni gifts, business corporations, the United Methodist Church, and other friends of the college.

In a discussion with Dr. Howell in mid-1981 he shared the cost savings incentive program that was in use at Central Methodist College. The request for ideas (Figures 13-1 and 13-2), how the plan works, and the first award are detailed in the following pages. Yes, there is a place for cost reduction in small business and the college environment.

Cost Savings Incentive Program
for Central Methodist College Employees

Purpose

1. To encourage all employees to become cost and energy conscious so as to realize the most efficient use of all our resources.
2. To encourage creative thinking and participation by making cash awards to employees submitting approved cost-effective

TO: ALL CMC FACULTY AND EMPLOYEES
FROM: THE PRESIDENT'S OFFICE
SUBJECT: CMC COST-EFFECTIVE SAVINGS PROGRAM

We are pleased to announce the implementation of a cost savings incentive program for Central Methodist College. This means you can receive cash for your money-saving ideas.

The procedure is simple. Look around and see if you can come up with an idea which will save us time, energy, and/or money. Write it down, talk it over with your supervisor or department chairman if you wish, submit your suggestion, sit back and wait for the action.

If your idea is accepted, you will receive a cash award. Your award will be 10 percent of the annual savings realized by your suggestion up to a maximum of $1000. An idea resulting in savings of more than $10,000 per year will be given special consideration.

All suggestions submitted will be reviewed to determine their feasibility. Determination of actual cost savings will be verified by the Comptroller.

The program begins immediately and will continue through June 30, 1981. At that time, the program will be evaluated. Your participation can make it work.

Why not take some time to jot down your idea on the attached suggestion form and send it to my office? Carol Mahaffie will collect the suggestions and get them to the appropriate people for review.

Good luck!

/cm
Attachment

Figure 13-1.   Request for ideas.

CMC COST SAVINGS INCENTIVE PROGRAM

1. SUBMIT SUGGESTIONS ONLY ON SUGGESTION FORM.
2. BE SURE FORM IS COMPLETE. IF YOU NEED ADDITIONAL ROOM, ATTACH ANOTHER SHEET OF PAPER.
3. IF YOU NEED ASSISTANCE IN COMPLETING YOUR IDEA, ASK YOUR DEPART-MENT CHAIRMAN OR SUPERVISOR TO HELP.
4. BE SURE TO SIGN YOUR NAME AT THE BOTTOM OF THE SUGGESTION FORM.
5. SEND YOUR COMPLETED SUGGESTION FORM TO THE PRESIDENT'S OFFICE.
6. ALL SUGGESTIONS WILL BE ACKNOWLEDGED.
7. ADDITIONAL FORMS ARE AVAILABLE FROM THE PRESIDENT'S OFFICE.

(a)

Figure 13-2. Suggestion form instructions (a), and suggestion form (b).

measures and giving them public recognition for their achievements.
3. To encourage budget chairmen to become more aware of how their dollars are spent.

Method

1. Suggestions are to be submitted in writing on approved form to designated person.
2. An evaluator is assigned to each suggestion. The evaluator will be an individual who has knowledge of the particular area covered by the suggestion.
3. The findings and recommendations of the evaluator shall be submitted to the comptroller who shall verify the actual cost savings. The suggestion will then be submitted to the president for final approval.
4. Awards shall be 10 percent (10%) of the annual savings up to a maximum of $1000. Special consideration shall be given to awards above the maximum level.
5. Final decision for special awards shall be determined by the president.
6. A token award of $10 shall be awarded for adopted suggestions where no direct cost savings result.

Responsibilities of the Supervisor

1. Encourage employees to submit constructive ideas.
2. Assist them in developing ideas when necessary and within reasonable limits.

Benefits

1. Reduces cost of the operation under his supervision through elimination of waste, improved handling, and better utilization of resources for overall effectiveness.
2. Improves safety through the correction of hazardous conditions.
3. Improves employee relations through a closer working arrangement with employees in helping them with their ideas and rewarding them for constructive thinking.

Responsibilities of the Evaluator

1. Investigate suggestions completely, promptly, and enthusiastically. Try to make something out of them.
2. Weigh pros and cons carefully and recommend adoption of the suggestion when the new way is better than the old. Be adequate in measuring benefits.
3. Discuss the results. Discuss rejected ideas with employees, explaining why their ideas cannot be adopted. Commend them for their efforts and encourage them to keep on trying.

Awards

1. Present awards to employees with complimentary remarks to gain the objective desired in submitting suggestions.
2. Winners should be recognized and rewarded as soon as possible. Procrastination, delay and confusion will cause the system to fail.
3. Acknowledge receipt of suggestions by putting lists of suggestions under consideration on bulletin boards, in employee publications, or by memo. As action is taken, the decisions are announced.
4. A good publicity program helps to maintain interest in the program. Included may be posters, news articles in local papers, letters from administration, etc.

Suggestion Earns $250 Bonus

Central Methodist College has implemented a cost savings incentive program for faculty, students, and staff employees. The program is

CENTRAL METHODIST COLLEGE
COST SAVINGS INCENTIVE PROGRAM

PROBLEM _____

_____

_____

_____

_____

_____

_____

SUGGESTION _____

_____

_____

_____

_____

_____

_____

EXPECTED RESULT _____

_____

_____

_____

_____

_____

NAME: _____   DATE: _____

DEPARTMENT: _____   SIGNATURE: _____

_____

_____

_____

Figure 13-2b.

designed to make everyone more aware of ways to help the college save money through energy conservation and reduced overhead.

Raymond Donley, inventory manager, received the first incentive award November 14. His suggestion for a way to clean the mopheads used by housekeepers will result in an immediate savings of more than $200 a month. On an annual basis this means a savings of about $2500 to Central.

Suggestions for savings are first submitted to the president's office for review. After a determination that the suggestion would be a financial savings in actual dollars spent, an award equaling 10 percent of the savings is given to the person making the suggestion.

For his suggestion, Donley received a check from CMC President Joe A. Howell for $250 and a certificate of appreciation.

Involvement, Opportunity, and Change

If you want to be successful in the quest for cost reduction, involvement is the name of the game. Two other essential ingredients are opportunity and the desire for change. Coupled with involvement, these provide an ideal environment for cost reduction.

My experiences with cost reduction have been at a number of Western Electric Company locations. The North Carolina Works, Winston-Salem, North Carolina, provided a solid foundation in the essentials of planning, developing, and follow-through that is required in any successful program. Working under the direction of two seasoned engineers, our challenge was to evaluate make-versus-buy decisions. To be more specific, we were working in the area of special packing materials for government products. Packing, shipping, and storage of the products turned into a unique cost reduction challenge.

The end result was that a special products shop for packing materials was established to support a wide cross section of production. The shop had the capacity to handle corrugated details, cartons, vapor bags, and exterior wooden shipping containers. Like most production operations the shop operated on a two-shift basis. As I look back, the major savings associated with the cases were in the areas of setup charges and small-volume unit pricing—paid to suppliers.

My experiences at the Kansas City Works became a unique challenge. As an industrial engineer, my main objective was to generate a part or subassembly through the manufacturing process in the most cost-effective manner. Providing coverage for a wage incentive group of 295 shop employees engaged in coil winding, varied assembly operations, testing, and rework of defective material was a full-time job. This environment provided a full-time challenge and also the opportunity to seek innovative cost reduction. In the high-volume operations small segments of time become extremely important. For instance, the start of a coil winding cycle, the actual count of finger movements, and the size of wire versus the size of hole in the slider mechanism can be very significant—when the volumes are large and 40 people are engaged in the operation.

Exposure to vacuum tubes, transistors, and other new technology devices produced at the Kansas City Works provided new and unique challenges in the ever-changing area of cost reduction. An increase in process yield of 1 percent, a small decrease in material usage, a slight reduction in test cycle parameters, a different saw blade for slicing wafers, all of these can and have been translated into cost reduction cases. An assignment in this type of production area can be a very rewarding opportunity that is there just waiting to be discovered. A teamwork approach works well—the production supervisor and engineer can make an unbeatable team when it comes to generating ideas. These ideas can then be transformed into decreased production cost. The end result—cost reduction savings.

In mid-June 1965, a fellow engineer mentioned an opening for an engineer at the Los Angeles Service Center. At this point in time I was not quite sure what the charter for a service center was. After a little investigation into the function of a service center I thought it sounded like an interesting assignment. In that time frame the service centers were provided engineering support from an engineering group at the headquarters locations in New York. However, the periodic engineering surveys, telephone discussions, and random visits from engineers left something to be desired.

With the help of my engineering supervisor, my resume for the opening was submitted for review by the Los Angeles management team. When the initial screening had weeded out a number of applicants, I was invited to Los Angeles for an interview. On the second day of my visit the production manager said, ''I think you are the engineer we have been looking for.'' After a positive reply on my part, he spoke in a very straightforward manner. One comment is still as fresh in my mind today as it was in September 1965.

The production manager outlined for me what he believed to be certain weak spots in his operation. After he concluded these remarks, he reached for his long-range planning calendar. The pages were turned to March 1966. He said, ''We are going to be reviewing your first six months' progress in mid-March 1966.'' With the calendar back in place he said, ''If you have not made a substantial contribution by then—you will find yourself back in Kansas City.''

A bit harsh you may ask? Not at all. When our discussion was concluded I knew what was expected of me. Also, I knew that what was or was not accomplished would be reviewed at a specified time. It was a pleasure to leave a meeting with a clear understanding of what was expected.

From this point on, the three elements of a successful cost reduction program were ever-present: involvement, opportunity, and change. My experiences at the Los Angeles Service Center were a challenge in every respect. Cost reduction was a new term. This does not imply that changes and improvements were not being made. The lacking element was documentation, which is essential to any cost reduction program.

The biggest challenge is explaining the new program to those who will be involved. This was accomplished and positive results were soon being achieved as a team effort.

With this background in mind, let us review a paper that was published in 1980. The intent was to cover background, and follow through in a sequential manner.

Cost reduction is related to problem solving. Some problems are known, others unknown.

A problem well stated is half solved.

John Dewey

COST REDUCTION—STEPPING STONE TO SURVIVAL IN THE 1980s*

Abstract

Cost reduction has been an effective tool utilized by Western Electric Company for over 50 years. Today engineers are coping with a dynamic technology and variable business conditions in order to remain competitive with general trade sources that manufacture similar end-use items for the telephone industry. This pressure will continue through the 1980s. Industrial engineers will continue to play an important role in combating this activity.

*By E. A. Criner, P. E., Senior Engineer, Western Electric Company, Sunnyvale, California 94086. Reprinted with permission from *Proceedings . . . 1980 Spring Annual Conference.* Copyright © American Institute of Industrial Engineers, Inc., 25 Technology Park/Atlanta, Norcross, Georgia 30092.

For engineers, the program serves as one of the yardsticks by which professional achievement is measured. For plant managers, the program is the key ingredient of interplant competition and the major means of holding price down. Reductions in direct labor and material have been and probably will continue to be the most fruitful areas. In 1978 a total of 6,389 cases were closed with savings of $225 million (1).

Introduction

"Saved: $32 million," in 1965

These bold headlines appeared in *Steel* magazine in August 1966.

A further expansion of the statement indicated that Western Electric Company, New York, was credited with these savings. These dollars were slashed from manufacturing costs.

In the previous five years, new cost reductions by the manufacturing and supply arm of the Bell Telephone System climbed from $14 million to $30 million annually. As a result of these savings, Western was able to lower prices of Bell System products by $44 million in 1964 and another $33 million in 1965 (2).

A total of 5,983 cost reduction cases were under way at the end of 1965. Approximately 1,000 cases were associated with major projects. A major case, at this point in time, was defined as one in which more than $2,000 was needed for development.

Background

Until 1966 almost all of the cost reduction dollars credited came from Western's 13 Works or plant locations. The new Service Division formed in 1962 was destined to become a major force in the growth of cost reduction in the Bell System. A vital link between Equipment Engineering, Installation Distribution, and Shop Repair, the Service Division was designed to bring Western closer to the Operating Telephone Companies. When the Service Division was completely formed, it was organized into seven geographical Regions in order to provide a "full service package" for the Operating Telephone Companies.

The need for qualified engineers to be resident at each of the 35 repair centers was soon recognized. This close association of technical methods specialists, supervision, and engineering was the start of a new wave of cost reduction. The Pacific Region, headquartered in Sunnyvale, California, has been a leader in cost reduction since March 1966. Five cases, with an estimated value of $132,000, were being developed at the Los Angeles Service Center in early 1966. The blending of engineering know-how from the Manufacturing Division coupled with the new frontier of the repair shop opportunity provided a fertile area for cost reduction development.

How the System Works

Engineers at all Western Electric locations can tap vital Company resources for data related to their specific projects. In general, good engineering—an evaluation of new materials, new methods, new developments, and new skills at the right time—characterizes most significant cost reductions. Also, other factors are good engineering, supervision, efficient group effort spearheaded by the Product Engineer, and cost-conscious Industrial Engineer.

Conducting a Cost Reduction Case is done in three distinct stages.

Opening the Case

After a review of considerations listed in Attachment 1 (3), the engineer documents his idea in an opening statement which is forwarded to the local Cost Reduction Committee for review and approval. See Attachment 2 for the format that is utilized. If the idea is beyond the scope of the local committee, all related paperwork is forwarded to the appropriate location for review and approval (4).

After the required approvals to open the case have been secured, the case is routed to accounting for checking and classification.

At this point the engineer is free to proceed with the development of his idea.

Conducting the Case

This is the phase in which the engineer has the opportunity to combine "blue sky concepts" with proven engineering principles. Development expenditures for equipment, design, and other related charges are made against the approved Cost Reduction Case. Cases can cover the gamut from a complete product design to a change in the packing carton or mode of shipment to the end use location.

Lapsed time in this phase can vary from several months to as much as five years or longer depending on the scope of the case. Collection of data and proving the idea is the challenge confronting each engineer involved in this endeavor.

ATTACHMENT 1

ENGINEERING COST REDUCTION ORGANIZATIONS

SUGGESTER

COST REDUCTION PROPOSAL (PCLF-409)

IDEAS

RECEIVE PROPOSAL

PRIOR CONSIDERATION — YES / NO

IS CASE FEASIBLE — YES / NO

ENGINEER ORIGINATES MINOR CASE (PINK SHEET)

ENGINEER CONDUCTS CASE AND ESTIMATES SAVINGS

IS THIS A MAJOR CASE — YES / NO

MAJOR CASE LIMITS
DEVT EXP 5K
PLANT EXP 6K
SAVINGS 15K

ENGINEER ORIGINATES MAJOR CASE (PINK SHEET)

DEPT. CHIEF ASST. MGR. APPROVAL

ENGINEER PRESENT CASE TO C/R COMM

C/R COMMITTEE APPROVAL — YES / NO

ENGINEER CONDUCTS CASE

IS CASE FEASIBLE — YES / NO

IS CASE FEASIBLE — YES / NO

MINOR CASE LIMITS EXCEEDED — YES / NO

ENGINEERS COMPLETES CASE
TIME STANDARDS
UPDATE INSTRUCTIONS
PURCHASE ORDER

ENG CALC SAVINGS CLOSING REPORT (GN 1786)

APPROVALS ENG, USER, ACCTG.

ACCTG ISSUES REPORT OF SAVINGS (M-577.1)

RECOGNITION TO CASE SUGGESTER

DEPT. CHIEF APPROVAL

ENGINEER CLOSES CASE

REPLY TO SUGGESTER

Attachment 13-1.   The processing of a cost reduction sase.

(⚹) Western Electric   ATTACHMENT 2

THIS
COPY
FOR

Investigation Case Authorization

35

CASE NO. _____
ISSUE NO. _____
PAGE _____ OF _____ PAGES

TITLE (BRIEF DESCRIPTION)

KEY WORD INDEX
1. _____
2. _____
3. _____

SCOPE AND OBJECT

LOCATION _____
ORGANIZATION NO. _____
PROBABLE COMPLETION DATE _____
CLASSIFICATION (SEE APPLICABLE INSTRUCTION)
☐ COST REDUCTION   ☐ DEFINITE ROUTINE
☐ OTHER DEVELOPMENT ☐ GENERAL ROUTINE
☐ BELL SYSTEM SAVINGS CASE
ASSOC. PLANT AUTH. NO. _____
DATE APPVD. BY C.R. COMM. _____
RATE OF RETURN _____
MAKE VS. BUY APPROVAL _____

OUTLINE OF WORK AND SCHEDULE

ESTIMATED COST AND SAVINGS DATA
1. DEVELOPMENT EXPENSE
   (A) ENGINEER'S SALARY LOADED   _____
   (B) EXPERIMENTAL SHOP WORK   _____
   (C) EXPERIMENTAL PLANT OR
       DEVELOPMENT FACILITIES   _____
                            TOTAL   _____
2. ASSOCIATED EXP. TO BE INCURRED
   AS A RESULT OF THIS CASE
   (A) MOVES AND REARRANGEMENTS   _____
   (B) REMODELING PLANT   _____
   (C) EXPENSE TOOLS AND SUPPLIES   _____
   (D) ENGRG. SERVICES   _____
   (E) SERVICES, OTHER THAN ENGRG.   _____
                            TOTAL   _____
3. NEW PLANT REQUIRED AS A RESULT OF
   THIS CASE (EXCLUDING ITEM 1 (C))
   (A) LAND, BLDGS., LAND IMPRV.,
       MACH. & TRANSP. EQUIP.
       (INCL. DESIGN)   _____
   (B) SMALL TOOLS   _____
   (C) FURN. AND FIXT.   _____
                            TOTAL   _____
4. TOTAL EXPENDITURES (ITEMS 1,2, & 3) _____
5. ANNUAL SAVINGS BASED ON
   (A) 5 YEAR AVERAGE   _____ _____
   (B) CURRENT LEVEL*   _____ _____
   COST ESTIMATE NO. _____ DATE _____

SUGGESTED BY   CONDUCTED BY

APPROVED BY: _____ ORG. NO. _____
* 12 MONTHS SUBSEQUENT TO THE DATE SAVINGS ARE MADE EFFECTIVE

	LAND, BLDGS. LAND IMPRV. MACH. & TRANSP. EQUIP.	SMALL TOOLS FURN. & FIXT.
6. PLANT TO BE REPLACED		
(A) COST	_____	_____
(B) COST OF DISPOSAL	_____	_____
(C) SALVAGE VALUE	_____	_____

REASON FOR REISSUE

APPROVALS:
PRELIMINARY _____ _____ _____

Attachment 13-2.   Format for opening statement.

Closing the Case

After the case has been investigated and the method of change has been implemented or installed, the case is prepared for closing. At this time a total economic package is presented to the local Cost Reduction Committee. Savings are expressed on a current level basis (the first year) and also on a five-year summary average.

Employee Recognition

After closing a Cost Reduction Case, employee recognition is a very important factor for each individual involved (5).

The Pacific Region has recently embarked upon a new program for employee recognition:

1. Publicity for adopted cost reductions in local newspapers and Regional news media.
2. Tangible rewards for cost reduction originators. These can range from a $25 gift certificate for cases with savings in the $5,000 to $25,000 range. A $50 gift certificate for savings in the range of $25,000 to $50,000. A gift certificate for $100 is awarded for savings of more than $50,000.

The above awards are presented at the close of the "Cost Reduction Year" by members of upper management. As an ongoing program, awards are made for cumulative savings on an individual basis. At a recent luncheon, seven engineers were honored. Two of the engineers had accumulated savings in excess of $1,000,000 and five others had accumulated savings exceeding $500,000.

How to Get Started

How do you set the stage to develop an idea that may develop into a creative cost reduction? Each individual may have their own unique plan that works for them. If you want to be an idea man, school yourself in these fundamentals:

1. *Think the "green-light, red-light" way*—Suppose your problem is lack of time to train workers. First, apply free-wheeling (green-light) thinking, turn up all the solutions: you can assign more workers . . . work longer . . . train on overtime . . . speed up training. Don't stop with less than 15. Then switch on your judgment (red-light thinking); pick the

best. Try to alternate your green- and red-light thinking at every stage from stating the problem to final "selling" of the answer.

2. *Narrow down the problem*—John Dewey, great American philosopher, said, "A problem well stated is half solved." To set up a target for your mind to throw snowballs at, you first have to decide what you want to be creative about.

3. *Concentrate*—If you worry over a hat full of problems all at once, you'll get flabby ideas. So put on your mental blinders. Concentrate on one—and don't let anything distract you.

4. *Keep plugging*—When you draw a blank, it is hard to keep on thinking. But Ten Eyck, Penn's great rowing coach, said, "If you can hold on just two strokes longer than your opponents, you'll win."

5. *Believe in yourself*—You can come up with good ideas. And you can do it in one or more of three ways: Imagination (weaving ideas into new combinations while thinking deliberately) . . . Inspiration (suddenly creating ideas automatically from chance observation or circumstance) . . . Illumination (letting ideas come out of nowhere, when you least expect them—after thinking has stopped).

6. *Let your unconscious take over*—When you're worn out, stop thinking. Your mind will keep on working even if you aren't consciously thinking. It will percolate fresh ideas when you return to your problem.

7. *Keep 'em flowing*—It's like a run of hot hands in poker. If your mind is hot now, keep thinking up ideas. If you stop, then try to start again later, you may be cold as a corpse.

8. *Act*—Idea creation begins with the hot flash. But it's not completed until you put your idea to work. Crude preparations of penicillin were described in 1929, but nobody followed through on the discovery for a dozen years.

Evaluating Results

After a general review of past history of cost reduction in the Bell System, the procedure for generating a Cost Reduction Case, and how new ideas are evolved, let's review some specific results.

It is clear that the year 1965 was a major turning point in our economy. All through the late Fifties and the first half of the Sixties inflation had averaged between 1.5 and 2 percent annually. The economy was strong, and—at least from the Bell System's standpoint—seemed able to accommodate some minimal level of inflation.

Even in those years the Bell System's productivity was increasing at more than twice the national average—enough for us to offset the higher costs that inflation of 2 percent or less each year imposed on us.

The effect was that the prices consumers paid for telephone service were stabilized.

Then, starting in 1965, inflation began to outpace our ability to keep up with it through productivity gains.

By 1967, three of the Bell Operating Companies found it necessary to ask their state regulatory commissions for the first general rate increases filed by any Bell companies in about 10 years.

Over the past decade, of course—with inflation surging to more than 11 percent in 1974 and above 13 percent in 1979—we have had to seek permission from the regulatory commissions to charge higher rates at numerous times.

The doubling of the general price level since 1967—going up more than 100 percent, in fact—has had an impact on our costs of providing service that can be illustrated by just a few examples.

The following charts illustrate the results that can be obtained through an effective Cost Reduction Program. Refer to Fig. 13A-1.

CHANGES FROM

1967 - 1978

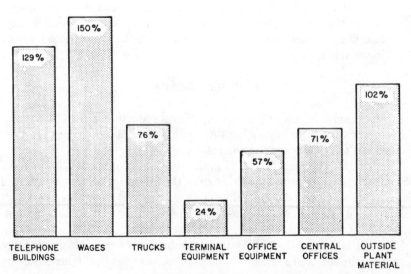

Figure 13A-1.   Price changes from 1967 to 1978. (*Source:* Western Electric Co.)

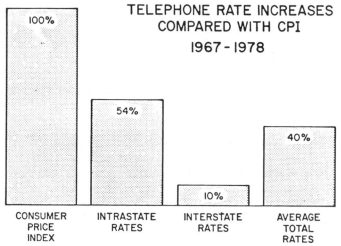

Figure 13A-2. Telephone rate increases compared with CPI, 1967–1978. (*Source:* Western Electric Co.)

Nonetheless, Bell System rates on the average have risen only 40 percent over the past 11 years—less than half as much as the Consumer Price Index. Refer to Fig. 13A-2.

That performance is based, first and foremost, on hard-won productivity gains that now are growing at three times the national average.

Much of our productivity growth flows from the Bell System's integrated structure which combines the roles of inventor, manufacturer, and operator in Bell Laboratories, Western Electric, and the Operating Telephone Companies, including Long Lines.

These Bell System partners work closely together, from concept to completion, on myriad projects to increase the quality and efficiency of high technology communications services. This teamwork creates an endless circle of concentrated effort that has blunted much of inflation's potential impact on our customer.

It is sufficient, I believe, to point out that the Bell System's technological breakthroughs have not only created productivity gains for us, but also for other companies which have built on our discoveries and developments.

For example, *Fortune* magazine (May 22, 1978, p. 108) has pointed out that of 18 basic advances in semiconductor technology, 12 have come from Bell Laboratories.

In addition to those discoveries, Bell Labs' invention of so-called "magnetic bubble memory" promises further huge advances in micro-electronic technology. Where we now are designing integrated circuits which put a whole computer with 65,000 bits of memory on one tiny chip of silicon, the application of "magnetic bubbles" to chip technology makes million-bit-memory micro-chips a very real possibility.

In brief, this ripple-effect of increased productivity has unquestionably served the nation well. Refer to Fig. 13A–3.

I might also mention the energy conservation route to productivity improvement that has been a deep concern of the Bell System since the oil embargo of 1973.

While our business has grown almost 50 percent since then, we are now consuming about 9 percent less energy. And the dollar savings are significant, indeed—more than $600 million saved since 1973, and more than $200 million in energy costs saved last year alone.

How have we accomplished this performance? By tight-fisted management. . . .

Now, turning from the most obvious impact of inflation on the Bell System—inescapable higher costs which force rates up despite our exceptional productivity gains—I want to cover more insidious results of the dollar's decline in value.

One of these results, of course, is the concept felt by Bell System employees—more than one million of them—who see the purchasing power of their income eroded by rising prices, despite the cost of living adjustments provided for in the contracts of those employees represented by unions.

But all of the productivity gains we can achieve by using our skills and technology and by cutting energy consumption—all of this is not enough to compensate fully for today's level of inflation.

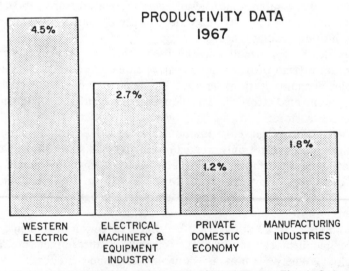

Figure 13A-3. Productivity data, 1967. (*Source:* Western Electric Co.)

Productivity improvement is—quite simply—the best way management in the private sector of the economy can try to hold down prices in the face of soaring costs (6).

Cost Reduction Examples

Over the past decade the cost of materials that Western uses to manufacture telephone and other related items has skyrocketed. Steel has gone up 213 percent in price, zinc 142 percent, copper 51 percent, and plastics 62 percent.

Labor cost over the same period rose 161 percent. Yet the cost of the typical telephone has risen only 18 percent (7).

How did Western manage this? The answer—through an effective cost reduction program.

Listed below is a cross-section of typical cases that have been closed recently.

1. Redesigned Telephone Set Base Plate.   One of the most analyzed parts in the telephone is the base plate. The part once painted is now plated. Sheet thickness has been reduced from 0.063 to 0.036 inch. In addition to plate thickness, four anti-skid feet were changed from suede to neoprene.

A recent update in the base is replacing the neoprene with an injection molded thermoplastic elastomer.

2. Redesigned Receiver Magnet.   This change involves a metal alloy used to manufacture magnets in the telephone receiver which connect electrical pulses into speech patterns. The introduction of the new material being utilized is "Chromindur" and will replace the present remalloy magnets. Cost savings of approximately $2.0 million a year are possible. In summary, production steps are reduced from 25 to 7.

3. Reusable Plastic Packing Details.   These details are used in conjunction with a wide variety of relay mounting plates. Molded from recycled ABS material, the details provide an economy in both material savings and labor requirements. Annual savings of approximately $55,000 per year.

4. Increased Recovery of Plastic Parts.   The base telephone is composed of four plastic parts—Handle, Housing, Receiver, and Transmitter Cap. Previous recovery programs such as buffing, small lot painting, were lacking in productivity and quality. Mismatched color between parts was a major problem. The introduction of painting on a larger scale—all component parts painted—led to cost savings and improved quality.

Introduced on a Regional basis, the savings were estimated to have savings

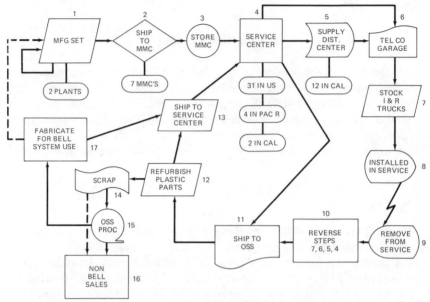

Figure 13A–4. Western Electric plastic recycle program.

potential of $500,000. Current national savings for this cost reduction are now in the range of $5.0 million per year (8). Refer to Fig. 4.

5. Recovery of U-Type Receiver Units. When the Varistor was damaged or the membrane was ruptured, the entire unit was junked. A repair procedure was introduced to salvage this unit. Original plans called for this unit to be recovered on a local basis. The concept was deemed to have high potential and was introduced for national application. Projected savings exceeded $1.0 million per year.

Summary

In the next ten years, Cost Reduction will play an even more important role in the struggle to stay competitive.

Cost reductions will continue to evolve through engineering involvement and constant review of What, Where, When, Why, How, and Who. These key questions are stepping-stones to survival in the 1980s. Soaring labor rates can be controlled to some extent. The larger gains in productivity will involve material and methods. These constraints will have a major impact on labor requirements—due to productivity gains. Productivity will continue to be the cost containment tool in the next decade.

References

1. F. J. Marcketta, Data Summary, Development and Cost Reduction Statistics Section.
2. *Steel* Magazine, The Penton Publishing Co., Cleveland, Ohio, August 15, 1966.
3. Cost Reduction Procedure Guide, Engineering Planning, Pacific Region, August 1979.
4. Cost Reduction Form, Investigation Case Authorization, GN-1780, October 1966.
5. E. A. Criner, Employee Recognition Proposal, Pacific Region, December 1967.
6. Newsbriefs Background, Western Electric Co., July 30, 1979.
7. Ernest Raia, Metalworking Editor, *Purchasing Magazine,* March 1979.
8. E. A. Criner, An Economic Analysis of a Plastic Recycling Program, *AIIE Proceeding,* 1977.

REFERENCE

1. Western Electric Co., *Newsbriefs Background,* November 9, 1981.

14
Cost Reduction Results

Western Electric's success in cost reduction has been and still is a major force in the fight against inflation. Results do not just happen; they are the cumulative effort of many people including managers, engineers, and others with input ideas. Remember cost reduction is an ongoing endeavor—it never stops.

In this chapter we will take a look at some of the challenges that dictate the need for an aggressive program. A cross section of cases will be reviewed. Also, a group of cost reduction cases in the form of a "brief" will be included.

THE ONGOING CHALLENGE

Each dollar earned or saved or spent in 1975 is now worth roughly two-thirds that much.* The money we try to save is getting harder to hold onto as prices for food, clothing, fuel, and other necessities seem to be rising faster than our income. Americans are more conscious than ever of the need to reduce costs, to economize, and to create ways to stretch dollars back to their former value.

The uncertain economy that has played havoc with personal finances has affected corporations in much the same way. The producer price index (once known as the wholesale price index) for industrial commodities has increased 60 percent since 1975. In particular, the cost of energy (fuels, related products, and power) for industrial users has increased 134 percent since 1975, and 473 percent since 1967.

What these statistics say is that it costs Western Electric a lot more for raw materials, supplies, components, fuels, chemicals, plastics,

---

*All statistics are from "Economic Report of the President," transmitted to Congress January 1981.

metals, and machinery. Quite simply, it is more important than ever to find and monitor cost-effective manufacturing techniques.

Western has been doing something about rising costs through the cost reduction program. Long before inflation became a household word—and worry—Western had its cost reduction system in place. Millions of dollars have been saved through this program over the years. It is one reason why the cost of telephone service has increased at a substantially slower rate than the cost of other consumer goods and services.

Western's cost reduction program—finding new and more economical ways of doing things—is an established part of doing business. Through it, results are measured, analyzed, and studied. The program itself costs money, and for that reason, its scope, size, and rigorous procedures are monitored closely.

Before any ideas for cost reductions can be fully investigated a feasibility study is conducted to estimate the savings and the expenditures for the coming five years. These estimates are compared to the out-of-pocket expenditures for the case to determine comparative profitability in order to choose projects with the greatest rate of return for the company's investment.

When a project that shows promise is found, a formal case is prepared and submitted to the local cost reduction committee by the engineers who proposed it. If it proves worthwhile, the idea is put into practice. The minutes of these committee meetings are circulated to all of Western's locations to avoid duplication of effort as well as to inform other engineers of possible application at other locations.

But what about quality? That is a major consideration in any cost reduction case. Many of the recent cost reductions are based on technological developments that help to save money and sometimes even enhance the product's versatility. The quality of Western's products is designed in at the beginning. The Labs must approve all changes made in the design of a product. In addition, each of Western's plants has a quality-assurance group that reports independently on the products. These controls make Western's one of the most sophisticated cost reduction programs in the industry.

In the 1980 Annual Report, as in past reports, a section was devoted to engineering cost reductions. Cost reductions in 1980, made possible by finding more efficient ways of producing products, resulted in about $295 million in first-year savings. That is a new record for the program. A small sampling of the hundreds of cases that made 1980 a record year are covered here. [1]

A CROSS SECTION OF CASES

Circuit-Pack Assembly at Montgomery

At the Montgomery Works, the assembly process for data set circuit packs has evolved from hand operations to very sophisticated computer-controlled machine operations that are almost fully automatic. This assembly process streamlines the product flow, uses computer-controlled automatic and semiautomatic machines, numerically controlled hand-assembly stations, automatic conveyors and wave-solder machines, numerically controlled lead-trimming machines, and an automatic defluxing machine that removes flux residue and cleans and lubricates the gold finger-type terminals.

Automatic assembly has resulted in a significant reduction in assembly defects. In 1980, a follow-up study was made to determine if the visual screening operation for components could be reduced or eliminated. After a careful review of the visual screening process as it was then carried out, the planning engineers for facilities and processes in data set circuit-pack manufacturing at Montgomery found that the visual screening process duplicated screening effort that was no longer required. Further, some of the screening that required partial disassembly of the circuit packs could be done prior to assembly.

The engineers developed new screening instructions for circuit-pack products that eliminated duplication and the need for disassembly. In some cases they found sample screening to be adequate. These cost reduction measures were expected to provide $1.0 million in savings in the first year.

Saving in Gold In Dallas

Here is a cost reduction case that actually occurred at Dallas, but is typical of the kind of thing going on at many Western Electric locations. When the price of gold began to climb several years ago, it became increasingly important to watch its use and to safeguard against the slightest waste.

At the suggestion of Bell Labs, four engineers from the Dallas Works presented a cost-saving idea that involved reducing the amount of gold plating on the contact and backplane ends of the 947K connector terminals from 125 microinches and 100 microinches, respectively, to 50 microinches. They reasoned that if they could reduce the gold requirement and still meet the design requirements and practical porosity specifications on the connectors, Western could save a considerable amount of gold purchases.

A microinch is one millionth of an inch. The microscopic amounts of gold represented by 50 or 75 microinches may not seem like a great deal—they are too small to be seen—but, in the process of producing millions of connectors each year, enormous savings can result.

After working with Bell Labs to test the connector's efficiency, the gold reduction on the contact end was approved and is now being implemented throughout Western. The current-level savings that resulted from this case totaled more than $2.5 million.

New Repeater Case for SLC*-96 System

While Western is primarily the manufacturing unit of the Bell System, the group at Springfield is responsible for engineering, quality, and marketing of products purchased from suppliers outside the Bell System. Western is one of the nation's largest consumers, purchasing over $5 billion in supplies each year. Engineers at Springfield are responsible for the design and development of purchased products and manufacturing processes at outside supplier facilities.

At the Purchased Product Engineering facility in Springfield, New Jersey, two engineers incorporated a new metals-casting technique on repeater cases and achieved a $5.4 million cost reduction. The year was 1976 and that case was the largest in PPE's history.

In 1980, four engineers working as a team achieved outstanding results for PPE. This time, the savings to Western totaled approximately $6.4 million, which represents the largest cost reduction at any of Western's locations in 1980. The new cost reduction is the result of the design of a new repeater case that is smaller, lighter, and better suited than the heavier cast-iron cases to the requirements of the subscriber loop carrier systems. Their new repeater case has a plastic housing for

---

*Trademark of Western Electric.

ease of manufacture and incorporates features that facilitate maintenance by the operating companies.

The newly designed repeater cases are to be used with the new SLC-96 system and other carrier systems using T1-type carrier equipment. These repeaters restore and amplify the multiplexed voice and data signals as they are transmitted over trunk cables. It is expected that approximately 10,000 cases will be manufactured on an annual basis.

Trimming Costs at Indianapolis

A cost saving of $1.3 million was realized at the Indianapolis Works in the assembly process of the Trimline telephone. The savings resulted from the design and installation of machines to carry out critical assembly processes that were intially done largely by hand. Two engineers designed a process and the necessary machinery to insert components into printed wiring boards in the Trimline phones. Although components have been inserted automatically in other printed wiring boards, the Trimline telephone network requires flexible boards, which have been considered unsuitable for automatic insertion.

The new machines at Indianapolis simultaneously cut, form, and insert six electrical components. The machines eliminate handling, conveyorize many of the operations, and produce higher-quality products that require fewer repairs at the end of the line. Cost reduction is the end result.

Greater Yields on Memory Chips at Allentown

At Allentown four engineers from the chip manufacture and packaging groups joined forces to maximize yields and reduce delicate hand work in the production of a 65,536-bit semiconductor memory device, known as the 34A Dynamic Random Access Memory (RAM).

These RAM chips are manufactured on wafers of silicon that have been chemically and thermally treated, and each wafer contains about 150 individual circuits when the processing has been completed. Because of the minute size of features on the chips and the complexity of the process, they are susceptible to defects introduced during the more than 130 different manufacturing operations.

As far as chip designs, several problems associated with the electrical performance and the chip layout were identified. Certain specific processes seemed to be a control factor that limited the yield of chips. Western and Bell Labs analyzed the production operation to pinpoint the specific factors that were limiting yields. The team then improved the manufacturability of the product and more than tripled the chip yield per wafer.

In the packaging area, semiautomatic package handlers were developed for the test set to allow one operator to run up to three sets at once. The Allentown engineers were able to improve productivity by establishing routines for fewer operators and by devising more reliable ways to test the chips. This cost reduction case resulted in a $9.0 million savings.

Western's cost reduction program is just one part of the effort to keep costs down and productivity up. Cost consciousness is more than counting dollars and cents. To Western's engineers, it is a dedicated pursuit for quality and the most productive way of manufacturing products. It is a combination of good thinking and careful planning that produces opportunities for reducing costs. It means better service and better products for Western's customers.[2]

Comments on Results

The cases cited from Montgomery, Dallas, Springfield, Indianapolis, and Allentown did not just happen. They are outstanding examples of cost reduction involvement. They all have one thing in common—each one started with an idea. All the ideas had one thing in common. The end objectives was to produce a quality product in a more cost-effective manner.

Let us review a few more examples. Case study example 3 deals with the application of an energy management technique at Paul Masson Vineyards Winery. A wide cross section of cost reduction briefs follows. The briefs cover a wide variety of applications.

CASE STUDY EXAMPLE 3

Case Description	Energy Management
Company	Paul Masson Vineyards Winery
	Saratoga, California

Data Source   Mr. Harvey Ornsdorf
                         Manager, Industrial Engineering
Estimated Savings $1.25 million/10 years

Description of the Paul Masson Site: Existing Facility

The Paul Masson bottling and wine processing plant is the only large industrial facility in the city of Saratoga. The site is approximately 25 acres, of which 18 acres are currently fully utilized. The location is at 13150 Saratoga Avenue, adjacent to the proposed Freeway 85 right-of-way. In addition to the approximately 250 employees, the facility is visited by approximately 250,000 tourists annually.

The Saratoga plant currently uses approximately 4,300,000 kilowatt hours (KWH) of electricity and 133,400 therms of natural gas. With cogeneration capacity of 315 KW, it is expected to use 1,400,000 KWH of purchased electricity and 288,000 therms of natural gas. Currently 15 million gallons of wine are processed and bottled annually at this facility.

Design and Site Layout: Future Plans

The addition of 100-percent bottling capacity and production in the next seven years will produce significant opportunities to use the process heat available from the recovery boiler. Electrical demand will likely reach 7,000,000 KWH based on refrigeration requirements and processing of 30 million gallons of wine. Any new plant additions would be planned to use absorption, chilling, and air conditioning as well as steam for space heating and hot water.

I. Definition of Problem

Energy costs are going up at a much more rapid rate than the general rate of inflation. A kilowatt hour cost approximately $.01 in 1972. It is projected to cost $.064 in 1980. This represents about a 20-percent compound growth rate. However, by far the largest rate of increase—about 50 percent—was experienced last year. The utility company's energy cost adjustment alone went up 136 percent. The price of natural gas has also increased rapidly, and its percentage rate of increase is predicted to be larger than that of electricity over the next decade. However, gas costs are now only about one-third the cost of electricity per million BTU's. In addition, the Public Utility Commission has mandated a lower (about 10 percent) gas price (Schedule G-55) for those customers who use gas to create electricity.

II. Proposed Solution

An opportunity exists to install a cogeneration system that will generate 1,670,000 KWH while satisfying 95 percent of the Saratoga plant's process needs for thermal energy. It is desirable although not necessary to convert the majority of the process heat needs to hot water from 100-psi steam.

The California Energy Commission gave Paul Masson a $7500 matching fund grant to study the feasibility of installing a topping cycle gas-fired engine generator and heat recovery system at the Saratoga plant. The results of that study are reflected in the capital project cash flow analysis attached. A sustained 290-KWH generator with a maximum output of about 315 KWH for up to eight hours per day is recommended. It is sized to handle the thermal load of the plant at double the current production volume through at least 30 million gallons of wine per year.

A hot-water system is recommended which will employ two 10,000-gallon storage tanks that are currently used for wine storage. A hot-water (200°F) return system will be used for makeup water when the boiler must be fired. This is anticipated for carafe capping and pasteurization only.

III. Expected Contribution

The system is expected to save $1,248,300 after-tax dollars during the first 10 years of production. With proper routine maintenance and major overhauls at approximately 25,000-hour intervals, the system should be productive for 25 + years. A salvage value of $60,000 at 10 years is anticipated because current systems up to 15 years old are commanding prices 50 percent of new costs. Energy conservation systems are expected to be even more important in future years, and costs may escalate in line with the greater savings then available.

A. Energy Cost Savings. First-year energy cost savings are $86,200. Ten-year before-tax energy savings are $1,893,700. These figures are based on electricity price increases of 17 percent for years 1–5 and 13 percent for years 6–10 of the project. Gas prices are expected to increase at 20 percent for years 1–5 and 15 percent for years 6–10. If electricity prices go up at 20 percent and gas at 25 percent as some economists project, the savings would be significantly greater. See Table 14–1.

B. Labor Savings. The facility's current high-pressure steam system requires a boiler mechanic to be on duty around the clock from 3 A.M. Monday to 3 A.M. Saturday. This need will be reduced to half-hour daily checks of the generator system and presence on initially a single shift when the carafe capper or pasteurizer is operational (currently about 150 days per year). Eventually the volume of the carafe is expected to increase sufficiently to require a second production shift. A conservative view of the labor savings was used for this evaluation. A savings of 1.5 shifts for the first five years and one shift per day for the next five years was used. The before-tax labor savings was calculated to be $44,700 the first year and $585,800 over ten years.

C. Other Savings. Although not figured in the cash flow analysis, other savings from the cogeneration project are significant.

This project allows an absolute increase in production volume of 100 percent without adding additional thermal process equipment. (The system also facilitates conversion to hot water.)

A 100-percent boiler makeup water return system at 200°F is possible with no additional cost other than approximately 50 feet of piping. (The savings on steam usage for the pasteurization and carafe capping was not taken into account.)

Time of use utility rate structure is too new to provide an experience factor. However, the proposed A-21 time of use rate charges are expected as follows:

A PERIOD	CAPACITY KW	DEMAND $/KW/MO.	SAVINGS $/MO.	ENERGY KWH	RATE $/KWH	SAVINGS $/MO.
On peak	315	1.11	$350	39,700	0.0216	$857
Partial peak	290	0.91	264	48,700	0.0163	794
Off peak	290	0.63	183	60,900	0.0123	
			$797	149,300		$2,400

Energy cost adjustments charge at $.04063/KWH	=	$6,066
Additional gas consumption $/month	=	$3,400
Fuel savings/month = ($6,066 – $3,400)	=	$2,666

Table 14-1.   Paul Masson—Saratoga Plant Estimated Annual Energy Costs.

FY YEAR	$/ANNUAL SAVINGS	$/KWH	KWH	$/ELECTRICITY SAVINGS	$/THERM COGENERATION	BUSINESS AS USUAL	& THERMS COGENERATION	&$ NAT. GAS
1981	86,200	0.075	1,670,000	125,300	0.482	0.533	91,000	39,100
1982	97,000	0.085	1,670,000	142,000	0.579	0.639	88,000	45,000
1983	120,400	0.103	1,670,000	172,000	0.695	0.767	85,000	51,600
1984	141,600	0.120	1,670,000	200,400	0.834	0.921	82,000	58,800
1985	166,300	0.140	1,670,000	233,800	1.000	1.100	79,000	67,500
1986	190,900	0.158	1,670,000	263,900	1.150	1.270	76,000	73,000
1987	220,000	0.179	1,670,000	298,900	1.320	1.460	73,000	78,900
1988	251,700	0.202	1,670,000	337,300	1.520	1.680	70,000	85,600
1989	287,900	0.228	1,670,000	380,800	1.750	1.930	67,000	92,900
1990	331,700	0.258	1,670,000	430,900	2.000	2.220	64,000	99,200
Total	1,893,700		16,700,000	2,585,000			775,000	691,600

Thus the generator also acts as a peaking power generator, cutting demand by an even greater amount during the peak cost and peak use time of day.

Cogeneration natural gas users will get priority on fuel in the event of low utility reserves because electricity is being generated with the gas. This will allow the plant to maintain nominal production (two to four bottling lines) by shutting down air conditioning and other nonessential electrical consumers during rolling blackouts. Even for periods of a few hours at a time, this would save the cost of sending the whole production crew home with a minimum of four hours' pay. It also guarantees the plant's ability to continue production during periods of low inventory should there be a prolonged electrical crisis.

The generator may be used on weekends for production or whenever required capacity for thermal and electrical warrants. These additional hours of operation would create additional savings beyond the 5760 hours/year assumed in the analysis.

Labor savings in addition to the boiler mechanics' routine operations may be realized. The maintenance of the new equipment is already accounted for in the economic evaluation. Reduced maintenance on the existing boilers was not considered. The fact that they go to standby status and are used on single shift when used instead of 24 hours per day will result in a lighter maintenance schedule, reduced consumption of chemicals, etc. Pumps, deaereator, boiler, and related equipment will not wear out as fast, and therefore a longer serviceable life and fewer overhauls will be required. These are actually offsetting savings from the maintenance cost of the cogeneration system.

Switchgear and other paralleling electrical equipment purchased for this project will be valuable as the facility expands operations. The cost of a second generator at a later date will be proportionally less than this initial project.

Energy and labor savings throughout the analysis were calculated on a conservative "most probable" basis. Other experts have predicted energy costs escalating at rates 15 percent and more above the rate of inflation. Labor savings were projected at an annual rate of 10 percent increase. Current inflation rates exceed that. The current labor contract is being negotiated. A 13-percent rate of increase for the next three years is probable.

IV. Other Factors to Consider

There is a potential savings of $20,000 on the purchase of the generator set and catalytic converter from at least one supplier if the facility agrees to a turnkey installation prior to July 31, 1980. Any delay beyond that date will likely result in a 6-percent increase in the capital cost. The difference is between an investment of about $172,000 and potentially $203,000 should the decision be delayed until the next fiscal year. This represents an 18-percent savings on the total project cost if prompt action is taken. This would more than pay the interest on the investment for the entire first year.

An additional consideration is that a turnkey construction project

Table 14-2. Thermal Energy Usage at Paul Masson.

EQUIPMENT	ENERGY SOURCE	ENERGY REQUIREMENT	BTU $\times 10^6$ TO EQUIP.	WINE PRODUCTION AND SCHEDULING
Pasteurizer	100 psig steam	800 lb/hr	15.2	1070 gal/hr (68–150°F), 16 hr/day, 18 days/year
Carafe capper	100 psig steam	300–500 lb/hr	4.6	130–150 line shifts/year, 10 hours/day
Heat exchanger				
28–60°F	130°F water	1200 gal/hr	6.4	3000 gal/hr (28–60°F), 8 hr/day, 15 days/year
45–60°F	130°F water	936 gal/hr	4.7	7000 gal/hr (45–60°F), 7 hr/day, 230 days/year
Clean in place	180°F water	1000 gal/day	0.6	230 days/year
Millipore filter	155°F water	300 gal/day	2.5	6–8 times/day, 230 days/hr
Wine transfer line	180°F water	1600 gal/day, Monday	1.9	45 days/year
		500 gal/day, other days	0.6	185 days/year
Bottling lines				
Line A	110°F water	300 gal/day	0.2	200 days/year
Line B	110°F water	300 gal/day	0.2	220 days/year
Line C	110°F water	300 gal/day	0.2	150 days/year
Line D	110°F water	200 gal/day	0.1	230 days/year
Line E	110°F water	300 gal/day	0.2	200 days/year
Line F	110°F water	300 gal/day	0.2	220 days/year
Cuvee	110°F water	300 gal/day	0.2	175 days/year
Bottling tanks	180°F water	3200 gal/day	3.2	230 days/year
Miscellaneous tanks	110°F water	400 gal/day	0.2	
Rest rooms	110°F water	300 gal/day	0.2	
Dishwasher	140°F water			
Office heating	180°F water	2400 gal/day	2.4	120 days/year

would lock in responsibility for successful installation and operation of the cogeneration system.

This opportunity is really only available to the facility now because it is virtually the first industrial plant in Northern California to be ready to install this type of cogeneration system. The technique is proven elsewhere and the equipment is readily available with reasonable lead times now. The supply situation is not expected to remain as responsive as more companies respond to the energy crisis and evaluate the alternatives to utility company supply as the sole source of industrial power.

The best available control technology for the emissions is relatively inexpensive and effective for the rich-burning otto-cycle gas engine that has been chosen. Sound attenuation is no problem with the hospital-type silencers selected. The location in the existing outdoor boiler room is safe and relatively inexpensive to the alternatives. The equipment will provide no increase in noise levels above that already created by the air compressor that is being relocated to make room for this installation. Piping and other auxiliary costs are minimized by the proposed configuration. See Table 14–2.

The public relations value of this project is significant. The winery will get positive media coverage and favorable local attention from its consuming public. Signs or displays on the public tour could highlight the energy-conscious, good corporate citizen image. The energy program and this project are already being promoted at engineering and energy-related seminars. The facility has been mentioned in the *Energy User News,* a national trade publication, for the award from the California State Energy Commission for the feasibility study.

V. Summary

The cogeneration or "total energy system" will continue to bear fruit into the next century. Now is the time to invest in future energy savings. A return of $1.25 million after taxes in the first 10 years is expected. Payback of 2.33 years and ROI of 50.9 percent make this project highly attractive. The project is an integral part of upgrading the thermal and electrical energy supply at the Saratoga plant and conversion from steam to hot water. Action should proceed immediately to take advantage of current equipment cost (savings of 18 percent) and to effect maximum energy and labor dollar savings.

A closing comment: A ceremony to dedicate the project was held on December 16, 1981. Governor Jerry Brown was the keynote speaker. He cited this achievement as one that other companies could follow and would follow as the energy spiral continued.

SIXTEEN VARIED COST REDUCTIONS

COST REDUCTION BRIEF

COMPANY: WILLIAM H. RORER, INC.

LOCATION: FT. WASHINGTON, PENNSYLVANIA

CASE DESCRIPTION: THIS CASE DEALS WITH THE DISPOSAL OF AN ESTIMATED 1,600 TONS OF WASTE PACKAGING MATERIALS. THIS MATERIAL IS CONVERTED INTO 13.5 BILLION BTUS OF HEAT.

ACTION REQUIRED: SAVNGS FROM USING THE SCRAP MATERIAL AS A FUEL IS FROM TWO SOURCES: (1) $85,000 SAVED ON CONSUMPTION OF OIL AND GAS. (2) A SAVINGS OF $19,000 BY REDUCING HAULING COSTS OF SCRAP MATERIAL. THE NEWLY INSTALLED TWO-STAGE INCINERATOR GENERATES ABOUT 7,000 BTU PER INPUT POUND. HEAT OUTPUT IS ESTIMATED AT 5.4 MILLION BTU PER HOUR.

SAVINGS: $104,000

COST TO IMPLEMENT: $275,000

RATE OF RETURN: 20 PERCENT PLUS INVESTMENT AND TAX CREDITS.

SOURCE: MATERIAL HANDLING ENGINEERING, AUGUST 1980.

COST REDUCTION BRIEF

COMPANY: WESTERN ELECTRIC COMPANY

LOCATION: SAN LEANDRO, CA (NCNSC)

CASE DESCRIPTION: INTRODUCTION OF AUTOMATED TAPE LABELING MACHINES IN THE WAREHOUSE AREA.

ACTION REQUIRED: THIS CASE DEALS WITH THE PURCHASE AND INSTALLATION OF THREE AUTOMATIC TAPE LABELING MACHINES. THE MACHINES MANUFACTURED BY 3M COMPANY, WILL REPLACE THE PRESENT MANUAL OPERATION FOR APPLYING REFLECTIVE TAPE ON CARTONS BEING DELIVERED TO THE SHIPPING DOCK.

SAVINGS: $138,450 FIRST-YEAR SAVINGS. THESE SAVINGS ARE A COMBINATION OF LABOR AND MATERIAL.

COST TO IMPLEMENT: $77,510. THIS INCLUDES DEVELOPMENT, ASSOCIATED EXPENSE, AND NEW PLANT.

RATE OF RETURN: 117 PERCENT

SOURCE: PACIFIC REGION COST REDUCTION AGENDA, MEETING #170, JULY 7, 1981, PAGE 5.

COST REDUCTION BRIEF

COMPANY: NORTH AMERICAN INDUSTRIES, INC.

LOCATION: WEINER, ARKANSAS

CASE DESCRIPTION: THIS CASE DEALS WITH THE APPLICATION OF ROBOT ARC WELDERS ON JIB HEADS FOR INDUSTRIAL CRANES. THE FILLET WELDS COMPLETED BY THE ROBOTS ARE MORE UNIFORM AND 40 PERCENT FASTER.

ACTION REQUIRED: A ROBOT WAS PURCHASED TO WELD A-37 MILD STEEL. THE LACK OF HIGHLY SOPHISTICATED PROGRAMMING IS A MAJOR ADVANTAGE WITH THE ROBOT POSITIONED AT THE WORK SITE, A WHEEL-TIPPED TEACHING HEAD IS SLIPPED OVER THE WELDING TORCH. THE WELDER-OPERATOR LEADS THE TEACHING HEAD OVER THE WELD AREA—THE MEMORY IS SET. IF THE FIRST WELD IS GOOD, SUBSEQUENT WELDS WILL BE IDENTICAL.

SAVINGS: $12,000 YEAR

COST TO IMPLEMENT: NOT STATED.

RATE OF RETURN: NOT STATED.

SOURCE: MODERN APPLICATIONS NEWS, JUNE 1981.

COST REDUCTION BRIEF

COMPANY: TEXTRON, INC.

LOCATION: PROVIDENCE, RHODE ISLAND

CASE DESCRIPTION: THIS CASE DEALS WITH THE APPLICATION OF ENERGY MANAGEMENT TECHNIQUES.

ACTION REQUIRED: A PROGRAMMABLE LIGHTING CONTROL SYSTEM WAS INSTALLED. COVERAGE WAS PROVIDED FOR AN OFFICE AND FACTORY AREA (160,000 SQUARE FEET). THE PRESENT SYSTEM IS USED TO CONTROL ONLY THE LIGHTING. FUTURE PLANS CALL FOR LINKING HEATING, AIR CONDITIONING, AND AIR COMPRESSORS INTO THE SAME SYSTEM.

SAVINGS: $130,000 YR. PROJECTED

COST TO IMPLEMENT: NOT STATED.

RATE OF RETURN: 218 PERCENT.

SOURCE: BUILDING MAGAZINE, JUNE 1981.

COST REDUCTION BRIEF

COMPANY: BANK OF AMERICA

LOCATION: PASADENA, CALIFORNIA

CASE DESCRIPTION: THIS CASE DEALS WITH A METHODS REVIEW ASSOCIATED WITH THE PROCESSING OF CHECKS.

ACTION REQUIRED: CHECKS DRAWN ON OTHER BANKS WERE ROUTINELY BE-ING HELD AT A BANK OF AMERICA CENTER IN PASADENA. THE CHECKS WERE HELD JUST IN CASE THE FIRST FILMING TURNED OUT TO BE DEFECTIVE. A CHANGE IN PROCEDURE WAS IMPLEMENTED — THE CHECKS WERE RELEASED AFTER FILMING. THIS CHANGE REDUCED B OF A'S DAILY FLOAT BY $4.8 MILLION. THE RESULT, AN IMPROVED FLOW OF CHECKS.

SAVINGS: $703,000

COST TO IMPLEMENT: NONE — A METHODS CHANGE ONLY.

RATE OF RETURN: NOT STATED.

SOURCE: SAN FRANCISCO EXAMINER, MAY 10, 1981.

COST REDUCTION BRIEF

COMPANY: BELL AND HOWELL COMPANY

LOCATION: CHICAGO, ILLINOIS

CASE DESCRIPTION: THIS CASE DEALS WITH THE STANDARDIZATION OF COM-PANY FORMS, ELIMINATION OF NON-ESSENTIAL FORMS, REDUCTION OF FORMS INVENTORY, LABOR SAVINGS, AND FLOOR SPACE SAVINGS. IN SUM-MARY, A PERFECT EXAMPLE OF THE INTERWOVEN PRINCIPLES OF COST REDUCTION.

ACTION REQUIRED: STANDARDIZED FORMS WERE REDUCED FROM 700 TO 350, SIZES OF STANDARD FORMS WERE REDUCED FROM 43 DOWN TO 18. COST SAV-INGS ON FORMS—$28,000. REDUCED INVENTORY PROVIDED A REDUCTION OF TWO RECORD CLERKS AND TWO OTHER FORMS EMPLOYEES. SALARY SAVINGS $50,000. A SAVINGS OF 4,000 SQUARE FEET OF FLOOR SPACE WAS ALSO ACHIEVED. AT $6.00 A SQUARE FOOT THIS AMOUNTED TO $24,000.

SAVINGS: TOTAL SAVINGS $102,000

COST TO IMPLEMENT: NOT STATED.

RATE OF RETURN: NOT STATED.

SOURCE: MOORE BUSINESS FORMS, GLENVIEW, ILLINOIS.

COST REDUCTION BRIEF

COMPANY: U.S. NAVY–NAVAL WEAPONS CONTROL CENTER

LOCATION: CRANE, INDIANA

CASE DESCRIPTION: THE NAVAL WEAPONS CONTROL CENTER HAS THE FUNC-
TION TO WAREHOUSE AND SHIP MORE THAN 250,000 HIGH VALUE SERIALIZED
ITEMS YEARLY. MAINTAINING AN UP-TO-DATE INVENTORY WAS A CONSTANT
CHALLENGE.

ACTION REQUIRED: AN UPDATED INVENTORY SYSTEM WAS NEEDED TO
REPLACE THE OLD METHOD OF PROCESSING INDIVIDUAL CARDS FOR EACH
ITEM. RECONCILIATION WAS A MAJOR PROBLEM. A NEW METHOD OF BAR
CODE LABELS WAS INSTALLED. SPECIAL LABELS ARE PREPARED FOR EACH
NEW ITEM. A PEN LIGHT SCANNER IS USED TO RECORD DATA. THE DATA IS
CONVERTED TO MAGNETIC TAPE FOR PROCESSING BY THE COMPUTER.

SAVINGS: $100,000 YEARLY

COST TO IMPLEMENT: ESTIMATED—$8,000.

RATE OF RETURN: ESTIMATED—725 PERCENT.

SOURCE: MODERN MATERIALS HANDLING, BAR CODED INVENTORY, FEBRU-
ARY 1980.

COST REDUCTION BRIEF

COMPANY: AMERICAN EXPRESS COMPANY

LOCATION: ATLANTA, GEORGIA

CASE DESCRIPTION: THIS CASE DEALS WITH THE USE OF A PROGRAMMABLE
ENERGY CONTROL SYSTEM.

ACTION REQUIRED: AN INITIAL ENERGY USAGE SURVEY WAS MADE TO
DETERMINE ENERGY USAGE FOR HEATING, LIGHTING, AND AIR CONDITION-
ING. AFTER A REVIEW OF THE STUDY A PROGRAMMABLE CONTROL SYSTEM
WAS INSTALLED.

SAVINGS: ESTIMATED $60,000 YEARLY.

COST TO IMPLEMENT: NOT STATED.

RATE OF RETURN: NOT STATED.

SOURCE: GENERAL ELECTRIC COMPANY, WARWICK, RHODE ISLAND.

COST REDCUTION BRIEF

COMPANY: PACIFIC SOUTHWEST AIRLINES

LOCATION: SAN DIEGO, CALIFORNIA

CASE DESCRIPTION: AN UPDATED LIGHTING SYSTEM IN THE MAINTENANCE HANGAR HAS IMPROVED WORKING CONDITIONS AND REDUCED COSTS. BASED ON $.10 PER KWH AND AN ESTIMATED 3,600 HOURS PER YEAR, SUBSTANTIAL SAVINGS AND A 50 PERCENT INCREASE IN LIGHTING LEVELS WERE ACHIEVED.

ACTION REQUIRED: ORIGINAL PLANS CALLED FOR A RELAMPING OF THE 64,000 SQUARE FOOT HANGAR, A TOTAL OF 288 NEW 400 WATT MERCURY VAPOR LAMPS. ESTIMATED COST ABOUT $8,000. A LOCAL UTILITY COMPANY RE-VIEWED ALTERNATIVES AND RECOMMENDED A TOTAL OF 144 HIGH PRESSURE SODIUM LAMPS. THE NEW SYSTEM IS MOUNTED 55 FEET ABOVE THE FLOOR ON 22 FOOT CENTERS. A LIGHT LEVEL OF 75 FOOT CANDLES IS MAINTAINED AT FLOOR LEVEL, 90 FOOT CANDLES IS MAINTAINED AT WORKING LEVEL OF THE 35 FOOT WORK STANDS.

SAVINGS: $48,000 PER YEAR.

COST TO IMPLEMENT: $20,000

RATE OF RETURN: PAY BACK PERIOD OF FIVE MONTHS.

SOURCE: GENERAL ELECTRIC COMPANY, VIENNA, VIRGINIA.

COST REDUCTION BRIEF

COMPANY: BOEING COMPANY

LOCATION: EVERETT, WASHINGTON

CASE DESCRIPTION: THIS CASE DEALS WITH AN ALTERNATE ENERGY SOURCE TO REPLACE THE OIL, NATURAL GAS, AND ELECTRICITY THAT WAS PRE-VIOUSLY CONSUMED. THIS APPLICATION IS DESIGNED TO PROVIDE UP TO 60 PERCENT OF ENERGY REQUIREMENTS.

ACTION REQUIRED: A NEW $3.0 MILLION BOILER AND TRASH-HANDLING SYSTEM, TO GENERATE STEAM FROM LUMBER, PLYWOOD, CARDBOARD, AND OTHER PLANT WASTE HAS BEEN INSTALLED. HEAT WILL BE GENERATED FOR A SIX MILLION SQUARE-FOOT BUILDING. THE PROJECT WAS FINANCED WITH TAX-FREE INDUSTRIAL REVENUE BONDS. WASTE-BURNING BOILERS ARE ANTI-POLLUTION DEVICES.

SAVINGS: $300,000 LANDFILL CHARGES PLUS FUEL SAVINGS.

COST TO IMPLEMENT: $3.0 MILLION

RATE OF RETURN: 35 PERCENT

SOURCE: "SPECIAL CONSIDERATIONS," BUILDING AND CONSTRUCTION," JUNE 1981.

COST REDUCTION BRIEF

COMPANY: GENERAL SIGNAL CORPORATION

LOCATION: REGINA VACUUM-CLEANER DIVISION

CASE DESCRIPTION: IMPROVEMENT OF THE SWITCH ASSEMBLY IN VACUUM CLEANER HANDLE AND OTHER ASSEMBLY TECHNIQUES. THESE METHODS CHANGES ALSO LED TO IMPROVED QUALITY LEVELS AND FEWER REJECTED LOTS.

ACTION REQUIRED: A MORE EXPENSIVE SWITCH WAS PROVEN IN FOR USE. THE NEW SWITCH AT TWICE THE COST REDUCED THE REJECTION RATE FROM 20 PERCENT DOWN TO LESS THAN 1 PERCENT. THIS COUPLED WITH OTHER PRODUCTIVITY IMPROVEMENTS TRIMMED THE COST OF ASSEMBLING A VACUUM CLEANER FROM $1.56 DOWN TO $.88 EACH. A SAVINGS OF $.68 PER UNIT WAS ACHIEVED.

SAVINGS: $12 MILLION DOLLARS — OVER A THREE-YEAR PERIOD.

COST TO IMPLEMENT: NOT STATED.

RATE OF RETURN: NOT STATED.

SOURCE: LUBAR, ROBERT, REDISCOVERING THE FACTORY, *FORTUNE MAGA-ZINE,* JULY 13, 1981.

COST REDUCTION BRIEF

COMPANY: LOCKHEED MISSILES AND SPACE COMPANY, INC.

LOCATION: SUNNYVALE, CALIFORNIA

CASE DESCRIPTION: THIS CASE DEALS WITH THE USE OF LASER SOLDERING TECHNIQUES ON PRINTED CIRCUIT BOARDS. THE OLD METHOD UTILIZED A MANUAL REFLOW SYSTEM AND ALLOWED SOLDERING ONE COMPONENT AT A TIME.

ACTION REQUIRED: A MICROSOLDERING SYSTEM WAS PURCHASED FROM APOLLO SYSTEMS, INC. THE AUTOMATIC LASER SYSTEM SOLDERS ONE FULL ROW OF LEADS AT A TIME. A SAVINGS OF $46 PER CIRCUIT BOARD WAS REALIZED. THE LASER METHOD GENERATES MINIMAL HEAT. QUALITY IS ALSO IMPROVED. A FULLY PROGRAMMABLE MICROPROCESSOR CONTROLS THE LASER.

SAVINGS: $552 PER SYSTEM PRODUCED

COST TO IMPLEMENT: $51,000 ESTIMATED

RATE OF RETURN: PAYBACK PERIOD EQUATED TO THE PRODUCTION OF 92 REMOTELY PILOTED VEHICLES.

SOURCE: LOCKHEED MSC STAR, MAY 22, 1981.

COST REDUCTION BRIEF

COMPANY: INTEL CORPORATION

LOCATION: SANTA CLARA, CALIFORNIA

CASE DESCRIPTION: THIS CASE DEALS WITH A NEW AND ONGOING APPROACH TO INCREASING PRODUCTIVITY. JOBS AND RELATED PROCEDURES ARE EXAMINED IN DETAIL TO REDUCE EXPENSES.

ACTION REQUIRED: THE PLAN FOR IMPROVING PRODUCTIVITY AND REDUCING EXPENSES BEGAN IN THE INVENTORY AND SUPPLY FUNCTIONS. ACCOUNTS PAYABLE SERVES AS ANOTHER GOOD EXAMPLE, STEPS WERE REDUCED FROM 25 TO 14. INITIAL EFFORTS BY THE COMPANY AND WOFAC COMPANY, A CONSULTING FIRM, LED A REDUCTION OF 122 EMPLOYEES, AND REDUCED OTHER EXPENSES.

SAVINGS: $1.9 MILLION

COST TO IMPLEMENT: $240,000 CONSULTING FEE

RATE OF RETURN: NOT STATED

SOURCE: MAIN, JEREMY, HOW TO BATTLE YOUR OWN BUREAUCRACY, *FORTUNE MAGAZINE,* JUNE 29, 1981.

COST REDUCTION BRIEF

COMPANY: HENREDON FURNITURE COMPANY

LOCATION: NORTH CAROLINA

CASE DESCRIPTION: THIS CASE DEALS WITH THE USE OF WOOD WASTE FROM MANUFACTURING OPERATIONS AS A SOURCE OF FUEL FOR THE BOILERS. THE CHALLENGE WAS TO KEEP THE FUEL SOURCE DRY AND IN A FUEL-READY CONDITION YEAR ROUND.

ACTION REQUIRED: THE PROBLEM WAS TO SELECT A STORAGE MODE. AN AIR-SUPPORTED WAREHOUSE AT ABOUT $2.85 A SQ. FT. VERSUS A METAL STRUCTURE AT APPROXIMATELY $7 A SQ. FT. WAS SELECTED. IN LOW HEAT DEMAND SEASONS, THE WASTE IS BLOWN IN THE AIR VIA A 5" TUBE FROM THE PLANT. IN THE HEATING SEASONS, A HOPPER IS USED TO FEED THE WASTE INTO A SILO FOR USE. THE AIR WAREHOUSE IS INFLATED BY THE USE OF A 10 HP BLOWER WITH 30" BLADE.

SAVINGS: $120,000 FIRST-YEAR SAVINGS.

COST TO IMPLEMENT: NOT STATED

RATE OF RETURN: NOT STATED

SOURCE: MODERN MATERIALS HANDLING, STORAGE AND WAREHOUSING, JUNE 19, 1981.

COST REDUCTION BRIEF

COMPANY: GULF FURNITURE COMPANY

LOCATION: MOBILE, ALABAMA

CASE DESCRIPTION: THIS CASE DEALS WITH A REDUCTION IN WAREHOUSE STAFF. ALSO, DAMAGE TO IN-PROCESS MERCHANDISE WAS REDUCED. BEFORE THE NEW STORAGE RACKS WERE INSTALLED, THREE FULL-TIME EMPLOYEES WERE ASSIGNED TO REPAIR WORK.

ACTION REQUIRED: AFTER A HIGH-RISE RACK SYSTEM WAS INSTALLED ONLY ONE REPAIR PERSON WAS REQUIRED. DAMAGED PRODUCT IN-PROCESS WAS

REDUCED FROM $25,000 TO AROUND $1,000. IMPROVED STORAGE AND RECALL PROCEDURES HAVE DECREASED INVENTORY BY APPROXIMATELY $500,000. THE NEW SYSTEM IMPROVES OVERALL MERCHANDISING AND INCREASED EFFICIENCY IN SELECTING.

SAVINGS: $100,000 ESTIMATED FIRST-YEAR SAVINGS.

COST TO IMPLEMENT: NOT STATED

RATE OF RETURN: NOT STATED

SOURCE: *MODERN MATERIALS HANDLING,* STORAGE AND WAREHOUSING, JUNE 19, 1982.

COST REDUCTION BRIEF

COMPANY: WESTERN ELECTRIC COMPANY

LOCATION: INDIANAPOLIS, INDIANA

CASE DESCRIPTION: REPLACE OVERSIZED SUPPLY FAN MOTORS WITH SMALLER ENERGY-EFFECIENT MOTORS.

ACTION REQUIRED: REPLACE 30 20-HP FAN MOTORS SUPPLYING MAKE-UP AIR TO BUILDING 30 WITH HIGH-EFFICIENCY MOTORS. THE NEW MOTORS WILL BE SIZED TO ACTUAL LOADS. THE REPLACEMENT MOTORS WILL BE A MIXTURE OF 10-HP AND 15-HP SIZES.

SAVINGS: $14,800 IN FIRST-YEAR SAVINGS.

COST TO IMPLEMENT: $28,000 REQUIRED. THIS COVERS DEVELOPMENT, ASSOCIATED EXPENSE, AND NEW PLANT.

RATE OF RETURN: 52 PERCENT

SOURCE: INDIANAPOLIS COST REDUCTION MINUTES.

CONCLUSION

There is need for a practical approach when developing a company plan for cost reduction. The need is for a flexible plan that encourages and allows a company to use its resources as effectively as possible. That means getting more out for what goes in; in other words, a higher ratio of output to input. A company cost reduction program should be a permanent, continuing effort to improve the productivity ratio of output/input. Stated another way, it should be a continuing effort to reduce unit cost and overhead expense.

With this in mind, what should you do to contribute to the cost reduction effort? The answer is simple. Come up with ideas, recommendations, and suggestions on how to reduce costs. Normally, it is easier to find small cost reduction ideas than to find big ones, but the sum of the small reductions can add up to large savings. There are many sources of potential cost reduction ideas: better utilization of manpower, materials, machines, money, methods, and time. By systematic analysis of these resources, ideas may be identified and pursued for a possible cost reduction development.

In order to help keep costs to the customer as low as possible and at the same time continue to offer improved service, the ongoing cost reduction program will play an increasingly significant role in the future.

REFERENCE

1. Western Electric Co., *Newsbriefs Background,* November 9, 1981.

15
Getting into the Act

As you probably are aware, the growth rate of America's industrial productivity slowed down in the 1970s and slowed to a halt in mid-1980. However, during this same period, productivity continued rising in most other industrial nations. Consequently, some foreign companies that could once survive only in their protected home markets can now compete with American industry everywhere—including our home markets here in the United States. And they are doing so with considerable success. America's sagging productivity has been matched by sagging profits in many of our key industries—steel, ships, motorcycles, automobiles, machine tools, and home electronics, to mention a few.

What makes this particularly hard to accept is that America showed much of the world how to become more productive. Many of the productivity techniques we devised have become weapons aimed against us on the battlefield of international competition. Today, it is not unusual to read more and more news stories about American companies that are fighting to survive. There have been so many such stories in recent months that it is easy to forget that the seeds of the crisis took roots years ago. Technically competent, highly aggressive foreign competition did not just appear on our shores early one morning in 1980. Rather, they were virtually invited into our markets by more than a decade of low capital investment; limited R&D spending; high labor demands; and too little management attention to productivity improvement. Accordingly, American industry faces a tough challenge in the 1980s—and some of our companies may not survive. Hopefully, more and more American companies are beginning to share a sense of urgency about this serious problem—and many are working hard to improve their quality and productivity.

PRODUCTIVITY THE KEY

In our country, we have the technology, the people, and the other resources to meet these economic challenges. What we lack is a total commitment and a well-synchronized strategy. It is my belief that a truly effective response to our economic problems—and particularly the competitive Japanese initiatives—requires a unified effort by key segments of American society. During the past 25 years, I have been concentrating a great deal of attention on the subject of productivity improvement. I have come to the conclusion that one of the key reasons for productivity problems in the United States is the quality of our industrial output. Producing more—at the expense of quality—is not the way to increase productivity. Doing something over because it was not done right the first time decreases efficiency, wastes money, and lowers productivity. The end result is an increased cost to the customer.

Quality and productivity go hand in hand. Simply stated, improved productivity and improved profitability are inevitable by-products of improved quality. As you may recall, during the 1950s and 1960s, the Japanese had the worldwide reputation for producing inferior products. It was common to see their products at fairs, carnivals and in novelty stores here in the United States. In those days, anything marked "Made in Japan" was automatically considered by the typical consumer to be of poor quality and low reliability. This is no longer true today. The real question to be asked is what really makes the difference. The work ethic of the Japanese worker is not found in many U.S. plants today—but this is not because of any differences in the workforce. Rather, the principal reason for the different work ethic is largely a lack of management attention to quality. What has been missing is leadership with commitment and involvement, emphasis, and training. There are several examples where businesses in the United States were faltering, and Japanese businessmen entered with the key ingredients—leadership, capital, and attitude—and turned them around. For example, Quasar, Sanyo, Sony, Honda, and others have had remarkable performance under Japanese management with American workers.

There is really no mysterious secret to their success. Quality and cost containment is made a companywide goal—and it permeates the entire operation, from the president to the production operation, in Japanese-managed American plants. We find a highly motivated workforce; stringent and clear engineering design requirements that emphasize quality; extensive training programs; participative management; emphasis on teamwork; and management insistence on high quality and cost containment. In much of U.S. industry, management treats workers the same as they did at the turn of the century. Virtually all planning is done by engineers and planning specialists in isolation from the manufacturing supervisors and workforce—and, as a result, quality and productivity have suffered. One thing is for sure, the American workforce is not the same today as it was in the early part of the century. Today's workers are more knowledgeable—and are better able to participate in the planning and management process.

A NEW BEGINNING

We cannot change the course of events that have taken place in the past. Hopefully, we can gain from previous mistakes that have been made. As we look toward changing the operations of yesterday, one thing is certain. People—the joint efforts of people at all levels of the workforce—will be the key that leads to change. The involvement of people led by a progressive management team is a must for the change that is needed to overcome the inefficiency that we have come to accept in varying degrees.

Progressive companies both large and small that encourage all employees at all levels to look for ways to reduce costs will survive the economic storm. Other companies that continue to operate only with directives from the top down will not be able to survive the economic storm. One of the most important lessons that should have been learned over the years is the importance of management change. Management must view themselves as "facilitators of cultural change" within their organizations. They must create an atmosphere that leads to positive attitudes within the workforce.

Management must by word as well as example support an ongoing cost reduction program. To become part of the company management system, the cost reduction program should not be a "new program" or a "tacked-on" effort. It should become a daily way of life.

Successful cost reduction programs have these characteristics:

1. Carefully planned.
2. Management initiated and supported.
3. Employee accepted.
4. Participated in by all employees.

As with the introduction of any new concept or change, however, a word of caution about timing is in order. In developing and implementing the program, move no faster than you gain understanding and acceptance. With understanding and acceptance comes commitment. Management commitment is essential, employee involvement is a must for the plan to function. Depending on the size of a company and resources available to it, a productivity improvement plan can range in complexity from very formal to very informal. Although there is no relationship between the degree of complexity and chances of success, there is a relationship between understanding and acceptance of a plan and the degree of its success. To encourage this understanding and acceptance, at all levels, the plan should start with an "awareness" phase. This awareness phase should use all the available means of company communications and reach everyone from the chairman to the employee in the office or on the shop floor. The program should be defined in terms that all employees are accustomed to hearing and working with.

In addition to increasing awareness, the cost reduction program should:

1. Collect and examine output and input data.
2. Select data and construct output/input measures.
3. Identify problem areas—plan action steps and goals.
4. Modify the use of resources—capital, material, human resources.
5. Monitor, report, and review progress.

In collecting and examining output and input data for the various company functions, avoid generating new data. Use data that is readily available and that is meaningful and acceptable to the employees involved in that company function.

PROFIT CENTER APPROACH

Output and input data is usually available for all profit centers in company functions. For example, in marketing, the sales dollars are available; in the accounting and purchasing function, numbers and dollar volume of purchases are available; and in production, units produced and other data are readily available.

Select data and construct output/input measurements for the company and for its various service functions. The measures should be meaningful and current. They should relate as much as possible to the operations concerned and to the employees involved. Measures can be constructed for practically every function in terms of output/input. If there is concern that a measurement is not entirely valid, it should be used anyway with the admission that it is not entirely valid and that a more valid measurement will be defined as the effort continues.

It is very important to start developing measurements. They serve as baselines from which to measure change over a period of time and will permit direct comparisons with the effectiveness of the competition. They also act as excellent catalysts for activity by focusing attention on problem areas. Refer to Figure 15-1.

Following this process, productivity measures can be constructed for all or part of the functions in a company—top to bottom and across all functions. Problem areas in a company can now be identified more clearly.

Establish steps and goals for productivity improvement by examining the historical trends of the output/input measurements for the company and its functions. This should be done for the past two, three, five years or more. If possible, compare these trends with those of the industry or the competitors. Identify and examine the problem areas that are indicated by the unfavorable trends in the output/input measurements. Chances are, most of these problem areas are already known to management and/or the employees. Whenever possible, seek employee participation in solving the problems in developing plans for improvement, and in setting goals. Consider problem areas as potential cost improvement areas.

Target Improvement Areas

Some of the potential improvement areas may be identified through the approach of asking six key questions.

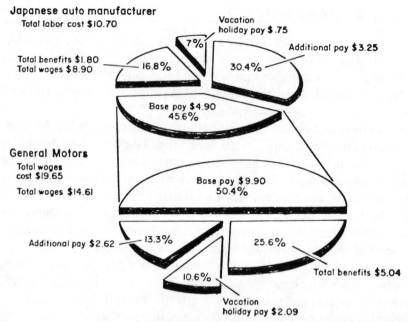

Figure 15-1.   Comparing hourly auto labor cost, Japanese automaker versus General Motors in the United States. (Comparison is based on third-quarter 1981 data.) (*Source:* General Motors.)

1. What is it?
2. Why do it at all?
3. Who should do it?
4. Where should it be done?
5. When should it be done?
6. How should it be done?

The effectiveness of a company will not improve, however, unless changes are made in the way resources are used. Changes can and must be made in the way a company uses its manpower, technology, and capital, and in the system itself in which resources are used.

The human resources of a company can be used more effectively through improved employee relations, involvement in developing solutions to problems affecting the work place, recognition, concern for job security, and the other factors contributing to job satisfaction. Particular attention might be given to the more effective use of employees in the "indirect" areas. Manpower is often referred to as the area of greatest untapped potential.

New technology in the industry or technology in other industries must be constantly examined as a possible means of improving productivity in the company. Improved technology should not be examined only as a means of improving productivity in the "direct" operations but as a means of improving the productivity of the "indirect" operations of the company, such as accounting and sales, as well. For example, the present generation of compact, easier-to-operate computers can be used in inventory control, accounting, distribution, and, in general, in the reduction of paperwork. There is a tendency to think of improvement through new technology in terms of the big breakthrough. It is important to recognize the little breakthroughs—the small improvements in technology—and take advantage of them. They can add up to sizable changes and appreciable cost reduction improvements.

The productivity of the entire economy is being adversely affected by insufficient capital being invested in plant and equipment. Most companies will invest in plant and equipment if it will improve their productivity and will give a reasonable return on their investment, even under the present tax laws and high interest rates. Although the long-range problem may be solved through more favorable tax laws and lower interest rates, the immediate problem is to identify changes in productivity that capital investment will permit and to make these investments. For example, an investment in compatible equipment that will reduce:

1. The costs of processing orders.
2. The maintenance and/or reduction of inventory of supplies, and inventory.
3. The costs of units of goods or services produced.

The entire company management system should be examined for possible changes that will improve the cost effectiveness of the company. It is important that a cost reduction program by companywide so that interdepartmental relationships are examined. A change that will improve operation in one department should not adversely affect the operation of another department or of the entire company.

Many of the problems in improving an industry's productivity are best solved by collective action. This is the area in which professional associations are well equipped to play a major role. This is particularly true in the measurement area. Industrywide data will provide interfirm comparisons. A company should not have to guess at its performance with respect to its industry, but should be able to measure itself against some valid data for that industry. For maximum improvement, each profit center should be examined as a system and the total company as a total entity.

When changes are made in the way resources are utilized, we must monitor, report, and review the progress of our improvement efforts. Cost reduction measurements and changes in these measurements should be integrated into the regular reporting system of a company. Improvements should be regularly reviewed with the employees involved. This is particularly important if the employees participated in developing the solutions and setting the goals.

The process of improving can be summarized as a process of measure, change, and review.

Measurement is extremely important for two basic reasons: It provides a baseline from which improvement can be measured, and, by helping focus attention on a problem, it becomes a catalyst for action.

Productivity improvement will come about as a result of changes in the way a company utilizes its manpower, capital, and technology. Unquestionably, a change in the marketplace and the economy will affect a company's profit picture. Productivity improvement enhances the opportunity for profit and is independent of changes in the marketplace.

Reviewing progress with employees not only lets them look at the score card but also provides them with the opportunity to become involved. Through participation, a company can take advantage of the know-how represented by the years of experience and expertise of all employees. Remember the expressions: "Your best consultants are already at work in your plant," and "Every pair of hands comes equipped with a brain." Encourage their involvement.

The goal of a well-planned and properly implemented productivity improvement plan is to integrate a process of change into the company's management system. This process of change will result in productivity improvement—the more effective use of all resources—from which everyone will benefit: owner, manager, supervisor, and worker.

HOW MUCH IS NOT ENOUGH?

Clearly, a suggestion or cost reduction proposal involving a product with high annual volume may have potential for large savings, but what about cost reduction proposals or suggestions involving low-volume products or operations? That is, how much total saving and, therefore, how many dollar savings per unit are required to cover the costs of investigation and development in addition to the costs of implementation of the idea?

Unfortunately, most suggestions from the typical manufacturing company suggestion system and many cost reduction proposals involve either too few units per year or inadequate savings per unit to pay for the cost of a detailed engineering department investigation.

In order to make the cost reduction and suggestion systems themselves productive, an efficient filtering mechanism is needed to separate the probable winners from the likely losers. The filter, however, must not throw out good moderate savings ideas, especially those with very low capital or tooling expense. In 1968, a concept called MINIPAY was introduced by C. W. Schilling in a small manufacturing plant in Indiana.

MINIPAY is a fast, simple, and effective remedy to purge the sure losers by comparing the normal investigation costs against each product's forecasted annual production rate, thereby determining the unit savings needed to break even. The minimum unit savings are calculated as follows:

1. Determine the minimum amount of potential savings in *dollars* necessary to justify engineering consideration of a proposal. This number may be found by multiplying the average engineer's loaded hourly salary by the average number of engineering hours required to investigate and implement typical small cases, divided by the cost reduction or suggestion system "batting average."

For example, assume that XYZ Company has a loaded rate of $40 per engineering hour, and the average simple suggestion requires 2.5 hours of engineering time to investigate and implement or reject. If the recent acceptance rate was 25 percent, the minimum savings required would be $400.

$$\frac{\$40}{\text{hour}} \times \frac{2.5 \text{ hour}}{\text{proposal}} \times \frac{1}{0.25} = \$400$$

If the company policy did not require the cost reduction or suggestion departments to be fully cost effective in that each proposal should stand only on its own merit, then the acceptance rate factor can be omitted. However, if the acceptance rate is very low, a program may be needed to educate the suggestors as to what constitutes a good suggestion.

2. Determine the weighted average hourly rate in the area affected by the proposal in *dollars per hour*. For our example, assume that Department 2 has an average total direct and indirect cost of $25 per direct labor hour. Further assume that Department 3 has a department cost of $80 per machine hour.

3. Determine the maximum payback period allowed in *years*. Payback period is frequently specified as a matter of company policy. In other cases, the payback period may be implied from the minimum acceptable rate of return or return on investment. For our example, XYZ Company required a payback in two years or less.

4. Determine the current or anticipated annual volume of each product in *units per year*. These numbers should be updated whenever the production forecasts are revised.

5. Calculate the minimum material or labor *dollar savings per unit:*

$$\underset{(\#1)}{\$ \text{ saving}} \times \underset{(\#3)}{\frac{1}{\text{years}}} \times \underset{(\#4)}{\frac{\text{year}}{\text{units}}} = \underset{(\#5)}{\frac{\$ \text{ saving}}{\text{unit}}}$$

6. Calculate the minimum labor *time savings per unit:*

$$\underset{(\#5)}{\frac{\$ \text{ saving}}{\text{unit}}} \times \underset{(\#2)}{\frac{\text{hours}}{\$ \text{ saved}}} = \underset{(\#6)}{\frac{\text{hours}}{\text{unit}}}$$

If the hours per unit is less than 1, convert the time savings to minutes or seconds per unit.

7. Tabulate the minimum dollars and labor savings required to repay investigation costs for each product as in the following example:

PRODUCT	FORECASTED ANNUAL PRODUCTION	MINIMUM $ SAVING/UNIT	MINIMUM HOURS SAVING/UNIT	
			DEPT. 2	DEPT. 3
A	2	$100.00	4.0 hr.	1.25
B	30	6.67	0.267(16 min)	—
J	575	.35	0.014(50.1 sec)	—
K	2,000	.10	0.004(14.4 sec)	—
L	10,700	.019	0.00075(2.7 sec)	—

Remember that these required savings are in addition to the costs of tooling or other process changes. Thus, from the MINIPAY table we see that a suggestion on product A to eliminate a 12-minute operation in Department 2 will obviously not repay the costs of a method change order as long as only two units are scheduled per year. In fact, the MINIPAY equations could be used to show that a minimum production of 40 units per year would be necessary to repay the average engineering costs to implement a 12-minute time saving in Department 2, given the problem assumptions. However, in Department 3, with much higher actual costs, an annual production of only seven units would repay the engineering implementation costs of a 12-minute savings per unit.

If the organization wants to adopt the change anyway to encourage the suggestion system, that is another matter. MINIPAY will quickly show the true costs of such a policy. Preferably, a standardized reply should be sent to the unsuccessful suggestor to the effect that, "Your suggestion and interest in improving the operation are appreciated. However, current production rates do not permit the suggested change."

Incidentally, we are not saying that significant time savings cannot occur on product A. On some products, such as large pressure vessels, time savings measured in hours are conceivable. The value of MINIPAY is to define the break-even point. On the other hand, look at product L in the example. Or better still, the methods analysts should be looking at L! A few seconds saved per unit could pay for some serious study on this product. Because the MINIPAY equations and supporting data are easily obtained either manually or by computer, tabulated MINIPAY information should be provided to all functional I.E. and methods personnel to pinpoint opportunities for truly significant savings as well as to eliminate the "no-go" losers.[1]

SELLING THE PROJECT TO MANAGEMENT

Many good ideas that have real cost reduction potential are never implemented for one simple reason. In many cases projects with real potential are rejected because the data presented to upper management is inadequate and incomplete. A proposal that has incomplete descriptions and inadequate economic justification is likely to be rejected. Projects that have been successfully installed have one common bond.

Before they were approved, a complete scope of work and economic justification was presented.

Engineering projects can be sold in an effective manner only if they are stated in the economic terms that management understands. There must be a clear understanding about a project before a decision can be made. With this key point in mind, think about the last idea you presented. What changes would you make if you had a second chance to sell the last idea that didn't quite make it?

In most cases, financial and operational managers do not have extensive engineering training. They make decisions about the expenditure of company funds. To perform this function effectively, they need a complete description and justification of a project before approving it. A statement from the engineer that he believes the project will work does not necessarily make it workable in management's eyes. Facts must substantiate beliefs. An organized, systematic approach that includes complete and detailed financial information makes the technical presentation more meaningful to management.

Every well-prepared cost reduction proposal contains two basic parts: the technical portion and the economic analysis. Proper preparation of these two sections ensures a better reception of the project, because management will find it easier to evaluate the project financially.

The organization of the material can also have an impact on selling an idea. There are two schools of thought on how to prepare an effective project report.

- One approach is to start with a general description, a statement of investigation, savings calculation, recommendations, and then the action summary.
- A second approach would be to start with a summary of the economic factors, recommendations, project overview, savings calculations, and any required appendixes.

THE REAL CHALLENGE

Develop and use the format that fits your needs and works in your environment. As a general guide for documentation, the following information should be prepared for each cost reduction project:

- Table of contents.
- Economic analysis summary.
- General description.
- Recommendations.
- Cost savings calculation.
- Financial information.
- Payback period calculations.
- Return on investment calculation.
- Five-year savings.
- Appendixes: technical data, miscellaneous.

1. The *table of contents* should be an index—brief, logical, and descriptive.

2. The *economic analysis summary* is critical to the overall presentation. It should contain all figures necessary for a quick analysis. When used with the table of contents, this summary should direct the reader to detailed and vital aspects of any project discussed in the presentation. It provides clear and basic financial information, the kind of data with which management is primarily concerned. If management needs further information, it can easily be obtained from the rest of the report.

3. The *general description* should outline what is required to implement the project and should review the pros and cons of the project. The possible effects of the project on existing operations should be evaluated. Articles in professional magazines or similar material that led to the interest in the item should be referenced. In large companies, if the project has application at other locations, this should be covered. Cite impact if possible.

4. The *recommendations* should be brief and the reasons for them should be explained. In this section, the positive aspects of the project should be emphasized. This section of the proposal is often one of the first to be read. Relate each recommendation to a target date for completion.

5. The *cost savings calculation* can be one of the simplest aspects of a project, or in some cases that involve a number of variables, it can be very complex. Keep in mind that this display is very critical in the development of any cost reduction pursuit. The types of savings that have to be calculated will depend on the complexity of the project.

Typical projects may involve labor savings, material savings, and expense savings.

6. The *financial information* section covers such subjects as cost of money, and the possibilities of leasing and direct borrowing. These items help present a better overview of the project. Two other items that may be included are the lease or loan period and the amount of the monthly payment, including principal and interest. Any other specific financial information that might be related to the project should be included.

7. The *payback period calculations* are used to substantiate the feasibility of the project. Although computing the payback period is simple, the figures that are used must be accurate. The following example illustrates the calculation, where: payback period in months = (project initiation cost, dollars) ÷ (total savings, dollars) × (duration of project, months). A project that requires $108,496 to implement is anticipated to save $690,238 over a five-year period. What is the payback period?

$$\frac{\$108,496}{\$690,238} \times 60 \text{ months} = 9.43 \text{ or } 10\text{-month payback}$$

8. The *return on investment calculation* is directly related to the payback calculations. Both sections of the presentation are important to management because they show how soon the project will pay for itself. The return on investment (ROI) can be calculated using the following equation: return on investment, percent = (period of operation, months) ÷ (payback period, months) × 100.

$$\text{ROI} = \frac{12}{9.43} \times 100 = 127.2\%$$

The return on investment is based on a 12-month period of operation. This figure indicates what proportion of the project will be paid for in the first year. These calculations will serve as a go–no-go gauge for each project. Management guidelines will vary from one company to another. In order to establish a screening technique it is essential to have established a lower limit that defines a return that is not acceptable. Each operation is different; however, a return of under 40 percent will probably be on a marginal side.

9. The *five-year savings* calculations are performed to show the total financial impact of the project under consideration. The calculations illustrate the long-term impact of the project. Refer to Figure 15–2.

The five-year data statement brings together on one sheet a summary of the following items:

- The net cost to implement the project includes all expenditures of funds that are required to transform the project into reality.
- Income tax credits are also shown, totaled by the proper category.
- Net cash return is shown for each of the five years. The net cash return is the summary of net savings plus income tax credits. In most cases it is anticipated that the net cash return will increase in each of the five-year periods. This is not always true since some applications may deal with a declining volume forecast. Higher savings may be obtained in the first, second, and third years. Reduced savings in the fourth and fifth may be the case.

This should be of no real concern in the evaluation phase. Just plug the numbers in and see what comes out. Total the net cash returns.

- Average annual return is the result of dividing the total net cash return by the number of years.
- Cost return ratio is obtained by dividing the net cost of the case by the average annual return.
- Rate of return for the project is also shown. As mentioned before, this is the final filter that determines if a project should be implemented, assuming that funds are available.

10. The final item to be included in the cost reduction project presentation is the *appendixes*. They should include all significant support material that substantiates the presentation. Among the items that should be included are manufacturers' and vendors' technical information; selected equipment performance data; drawings, schematics, technical diagrams, and graphs that explain the project; testimonial and reference letters about the selected equipment and its performance; terms and conditions; and cost information or bids ob-

```
                    RATE OF RETURN WORKSHEET
CASE NUMBER                                  OPENING
*****************************************************************

                    NET COST        TAX CREDIT
                    OF CASE      RATE      AMOUNT
-----------------------------------------------------------------

1 BUILDINGS & LAND           0.
2 OTHER PLANT            10000.    .0860        860.
3 INV. TAX CREDIT            0.    .0667        667.
4 ASSOC EXPENSE          11000.    .4300       4730.
5 DEVLP. EXPENSE          4132.    .4300       1777.
6 PRODUCT INVENTORY          0.
7 NET SALVAGE                0.
8 TOTAL                  25132.
-----------------------------------------------------------------



YEAR    NET SAVINGS        TAX CREDIT          NET CASH
                       PLANT     EXPENSE        RETURN
-----------------------------------------------------------------


1         13533.         860.     7173.        21566.
2         13533.         860.                  14393.
3         13533.         860.                  14393.
4         13533.         860.                  14393.
5         13533.         860.                  14393.
TOTAL                                          79138.

*****************************************************************
AVERAGE ANNUAL RETURN   $   15828.

COST RETURN RATIO       1.587857

RATE OF RETURN          63.5%


*****************************************************************


PREPARED BY: E A CRINER                      3-Jun-81


*****************************************************************
```

Figure 15-2.   Rate of return worksheet.

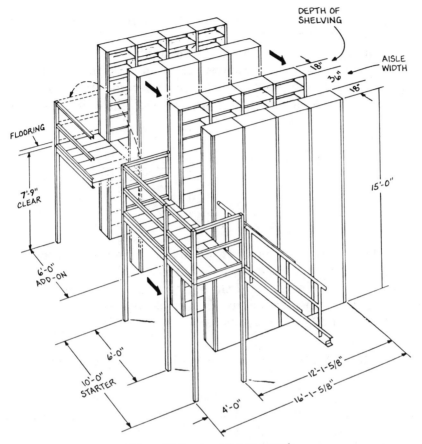

Figure 15-3. A warehouse example.

tained from manufacturers, vendors, or contractors. A tentative schedule for the project should be included. Refer to Figure 15-3.

Making projects a reality requires that they be approved by top management who have been convinced of their worth through a well documented presentation.

THE FINAL QUESTION

In most cases the final question to be asked and discussed is what is the minimum rate of return that must be achieved in order to assure management that the project will stand on its own? Stated another way, management needs to know the rate of return that is required to justify an expenditure of funds.

In order to answer this question, several inputs are necessary. The company tax bracket and the current cost of money (interest rate) are required. As an example, a company in 43-percent tax bracket confronted with a 20-percent interest rate may view the problem in these terms.

$$\text{min. rate of return} = \frac{\text{cost of money}(\%)}{\text{after-tax profit}(\%)}$$

$$35\% = \frac{20}{57} \times 100\%$$

The 35% is the minimum rate of return that is required just to break even. This minimum should be adjusted to include a profit objective and a risk factor. The adjusted rate of return, the sum of minimum return plus profit objective and risk factor, may place your requirement in the range of 50 percent or above.

$$\text{adjusted rate} = \text{min. rate} + \text{profit obj.} + \text{risk factor}$$

One thing to keep in mind is that these are dynamic numbers and will be subject to the current economic conditions. This factor coupled with your particular industry segment will determine which projects should be adopted and which ones should be dropped or held until a later date.

SOME VARIED EXAMPLES

Included for review are seven different ideas pertaining to effective ways to reduce costs. Each of these ideas presents a cost reduction potential. Although each one is different, they still have one common link. The link is that management had to be sold on the merits of making a change. The idea, the potential, and the savings can never be developed without a certain amount of behind-the-scene development before the management presentation. Develop your own style that works for you. Change the style from time to time in order to avoid falling into a rut.

COST
REDUCTION
POTENTIAL

APPLICATION:

LABOR . . . . . ●

MATERIALS . . ●

CAPITAL . . . . ○

ENERGY . . . . ○

EXPENSE . . . ●

IDEA: Investigate new ways to stabilize loads on pallets, reduce damage in shipping and storage.

APPLICATION: The use of a horizontal strapping machine was incorporated into the main conveyor and automatic palletizing system. This addition provided a way to improve stability of unit loads. The strapping is applied automatically by a sig node automatic machine. This system replaces a manual hand operation using heavy twine.

SAVINGS: The new method of securing loads saves approximately $30,000 yearly. Prior to this change, two employees were required to perform this function. Since the loads are now more stable, they can be stacked four high versus three high. This is another plus factor since it conserves on warehouse space.

SOURCE: Paul Masson Vineyards, Saratoga, CA

COST
REDUCTION
POTENTIAL

APPLICATION:

LABOR . . . . . ○
MATERIALS . . ○
CAPITAL . . . . ○
ENERGY . . . . ●
EXPENSE . . . ●

IDEA: A typical city government decides to investigate the required amount of street lighting versus associated cost factors.

APPLICATION: The conversion from mercury lamps to high-pressure sodium lamps on residential streets was evaluated. A relamping of 2,200 70-watt HPS lamps was substituted for an equal number of 175-watt mercury lamps. This is Phase 1 of a long-term project.

SAVINGS: A projected 750,000 Kilowatt hours of energy will be saved annually. A slight loss in illumination was noted. Annual savings of $75,000 with a payback of two and one-half years are anticipated. Light levels of approximately .35 foot candles are provided.

SOURCE: Maintenance Department, Ontario, CA

COST
REDUCTION
POTENTIAL

APPLICATION:

LABOR . . . . . ●

MATERIALS . . ○

CAPITAL . . . . ○

ENERGY . . . . ○

EXPENSE . . . ●

IDEA: Investigate the application of part-time or temporary employees for use
in the office, warehouse, and production areas.

APPLICATION: The use of part-time employees can be a cost effective mode of
operation. One large Silicon Valley company utilizes temporary people for up to
a maximum of 26 weeks at a time. These employees cannot be rehired for a period
of 18 months.

SAVINGS: Substantial savings can be accrued in direct labor as well as fringe
benefits. Yearly savings in the range of $9,000/$12,000 per year per employee is
possible. This method also provides a screening device for selecting permanent
employees.

SOURCE: A large Bay Area personnel firm.

COST
REDUCTION
POTENTIAL

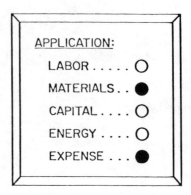

APPLICATION:

LABOR . . . . . ○

MATERIALS . . ●

CAPITAL . . . . ○

ENERGY . . . . ○

EXPENSE . . . ●

IDEA: Look for ways to reduce intransit damage as the product is moved from one point to another in the distribution cycle.

APPLICATION: Michigan Birch Door ships more than one million doors in their own fleet of nine tractor-trailers. Even with normal shipping precautions, damage to the product was running around two percent. On an order of 1,000 doors the loss of units valued at $100 each could run as high as $2,000 per load delivered. A methods review indicated that solid unitized load was required.

SAVINGS: Proper stacking to insure a uniform weight distribution was recognized as a must. Further study indicated that the strapping used to secure the loads was a key element. The change to a polyester strapping was instrumental in saving over $100,000 yearly.

SOURCE: Michigan Birch Door Company, Mt. Clemens, MI

COST
REDUCTION
POTENTIAL

APPLICATION:

LABOR . . . . . O
MATERIALS . . O
CAPITAL . . . . O
ENERGY . . . . ●
EXPENSE . . . ●

IDEA: Evaluate an energy recovery system. Ask one question - does your operation generate scrap paper, corrugated board, or other related material?

APPLICATION: The above question was asked at the Lockheed Electronics Plant in Plainfield, NJ. Their new energy recovery system is planned to save about 16 million cu. ft. of natural gas each year. The system is used to heat a 310,000 sq. ft. building. The system operates 10 hours daily, 5 days a week. The system replaces a gas-fired boiler.

SAVINGS: Yearly savings in the range of $119,000 are expected. This includes reduced energy consumption and trash hauling costs. A phase 2 change to burn liquid chemicals is planned. A third phase will use heat energy to provide air conditioning.

SOURCE: Lockheed MSC Star, May 8, 1981

COST
REDUCTION
POTENTIAL

APPLICATION:

LABOR . . . . . O

MATERIALS . . O

CAPITAL . . . . O

ENERGY . . . . O

EXPENSE . . . ●

IDEA: Evaluate shipping modes for letters, parcels, packages, and related bulk shipments up to 70 pounds.

APPLICATION: A catalog company that ships approximately 12,000 parcels weekly installed an electronic weighing system. The new scale replaced a fan scale used to calculate postage. A parallax error usually resulted in a over-statement of postage that was due.

SAVINGS: A major advantage of the new scale is the built-in "automatic shopper" This feature determines the least expensive mode. The cost of a one pount parcel shipped to zone 5 via parcel post cost $1.88 versus $.98 by United Parcel Service. Annual savings exceed $25,000.

SOURCE: Merlite Industries Inc., New York, NY

COST:
REDUCTION
POTENTIAL

APPLICATION:

LABOR . . . . . ○

MATERIALS . . ○

CAPITAL . . . . ○

ENERGY . . . . ●

EXPENSE . . . ○

IDEA: Conduct an energy audit of your facility. Evaluate ways to conserve electricity and natural gas.

APPLICATION: Evaluate the use of a computerized energy management system. Such a system can aid in reducing the quantity of electricity and natural gas usage. This can be achieved by effective utilization of heating, air conditioning, and ventilating. This technique was applied in nine buildings that were heavy energy users.

SAVINGS: The energy management system performs start-stop functions, and demand limiting on nine large air conditioning chillers ranging in size from 30 to 100 tons and other related equipment. Installation cost was $72,000 vs. $77,000 in projected first-year energy savings. Payback less than one year.

SOURCE: Converse College, Spartanburg, South Carolina

A CHECKLIST FOR PLANNING COST REDUCTION[*]

Y	N	R	
			Design Modifications
			1. Can a cheaper part be substituted for a more expensive one without loss of proper apparatus functioning?
			2. Can standard commercial parts be substituted for own-design parts, or vice versa?
			3. Can low-cost, high-production parts be substituted for high-cost, low-production parts?
			4. Can a part be combined with some other part, or can the part be eliminated by transferring its function to another part?
			5. Can specified tolerances be broadened to eliminate, reduce, or speed up existing operations or to reduce defectives?
			6. Can apparatus design be modified to adapt the item to assembly with a fixture rather than by hand operations?
			7. Can the item be redesigned to make it smaller or lighter?
			8. Can the use of small parts such as screws, washers, or rivets be eliminated or reduced?
			Manufacturing Methods
			1. Can a part be purchased more economically than it can be made, or vice versa?
			2. Can manufacturing operations be simplified, combined, or eliminated?
			3. Can assembly methods be speeded up—for example, with the use of power tools in place of hand tools?
			4. Can simple tools, fixtures, or jigs benefit production?

[*]Courtesy of *The WE Engineer.* Reprinted with permission from *The Western Electric Engineer,* October 1978, Cost Reduction Checklist, by Bob Zingali. Copyright © Western Electric Company, Inc. New York, N.Y.

Y	N	R

5. Can manufacturing methods be changed to reduce shrinkage losses or defectives?
6. Can material flow in the shop be improved?
7. Can the floor layout be changed to permit one operator to handle several machines?
8. Can a part be cast or molded rather than machined?
9. Can the cost of expense supplies be reduced?
10. Can a job be restructured to lower the labor grade required?
11. Can short-cycle assembly and wiring techniques be introduced?
12. Can visual aids facilitate assembly and wiring operations?
13. Can a conveyor be introduced to minimize handling and improve product flow?
14. Can work position layouts be improved to provide better time and motion economies?

Raw Materials

1. Are the physical, chemical, or electrical characteristics of the material out of which a part is made necessary to the part in its destined application?
2. Are the specified limits on the material unnecessarily close?
3. Can a cheaper material be substituted?
4. Can a cheaper finish be specified?
5. Can a standard-length material be specified to replace a special-length material?
6. Can outside suppliers' tooling be improved to reduce the cost of purchased components?
7. Can material procurement specifications be revised to effect reductions in inspection or associated costs?

Test Methods and Facilities

1. Can the introduction of more advanced test sets reduce test time?

Y	N	R

2. Can the use of go–no-go devices or digital meters reduce the time required to read meters?
3. Can a test fixture be applied to free the tester's hands for other tasks?
4. Can existing test fixtures be simplified by using magnetic devices, light sensors, speed clamps, etc.?
5. Can statistical sampling inspection replace 100-percent inspection?
6. Can test or inspection operations be eliminated, reduced, or combined?
7. Can the use of explicit troubleshooting instructions reduce test time?

Machine Equipment

1. Does the manufacturing tolerance in a piece part require the machine used to make the part?
2. Can design changes in the part reduce the cost of machining it?
3. Is a faster or less accurate machine suitable for making the part?
4. Can the part be made on a standard automatic machine?
5. Is the scrap produced by the machine excessive?
6. Can the scrap be utilized?
7. Can new features be added to the machine to improve its operation?
8. Can the use of new types of punches, dies, drills, taps, location devices, etc., result in greater output, longer life, or easier production?
9. Can parts be fed to the tool automatically?
10. Can a punch and die set economically perform several operations currently performed separately?
11. Can machining maintenance costs be reduced?
12. Can a new machine be justified to replace an old one that requires excessive maintenance or produces a high percentage of defects?

Lastly, attitudes *are* important; they do contribute to success. Specifically, it is recommended that you:

- Be inquisitive. Take nothing for granted. Last year's lowest cost may be this year's overpriced item.
- Be aware—aware of your product and its end uses, and aware of competition, both external and internal.
- Be defensive. Look at your job through other people's eyes. What would the new guy come in and see that you don't?
- Be persistent in seeking out ideas. Look at the process or product from every possible angle—and then look again. Don't fear failure. You have to have some duds, and nobody expects you to have a winner every time. It's like prospecting; if you keep looking all the time, you'll get lucky some of the time.
- Don't play it "close to the vest" with your associates. Share your ideas with them, and they will share their ideas with you.
- Remember shop supervision, purchasing, etc. They are all part of the team and can make valuable contributions.
- Be optimistic. There is a solution to every problem, and imagination, skill, and initiative are your stock in trade.

Cost reduction is more than an economic challenge to the company. It is a personal challenge to the engineer, and successful participation in the cost reduction program provides a real measure of self-satisfaction, in all, cost reduction is everybody's business. Make it your business.

REFERENCE

1. Schilling, C. W. Consulting Engineer, Indianapolis, Indiana.

16
Summary

INCREASED PRODUCTIVITY A MUST

In these times of continuing inflation and high interest rates, nothing is more important to the success of our businesses, to the standards of living that we want for our families and ourselves, or to the economic strength of our country than our ability to increase productivity.

In the quest to achieve increased productivity, cost reduction has been proven to be our most effective tool. Do not be misled when the term cost reduction is used. We must not delude ourselves into producing a product or service that offers only a lower price. If we do this, two other basic requirements have been overlooked.

A successful ongoing cost reduction program will produce a product or service that is competitive on the basis of price, quality, and service. If you have doubts on the validity of this statement you may want to examine the balance of trade payments between the United States and other countries. What does this have to do with productivity, you may ask? A valid question indeed.

The balance of trade could in fact be an indicator of productivity even if other more definitive measures were not available. Let us explore this idea.

A CASE IN POINT

Let us conduct a simple evaluation. Take two rows of parked cars in a shopping center parking lot. Count the total cars that are parked. Then count the number of foreign imports. What is the percentage of imports? The ratio may range from 15 to 40 percent or above in some sections of the country. Going one step further, conduct this same exercise in the reserved parking area at your work location. This can be an eye opener. Do not be surprised at the outcome.

Keep in mind that each individual who purchased a foreign import did so on the basis of price, quality, and service. These are in fact that same factors that are the basis for any cost reduction program.

A product in today's market plan must be able to compete with a wide range of competitive choices that are produced in this country as well as abroad. No product, company, or market segment is immune from the decision-making process that takes place before a product or service is purchased.

Before moving on, let me suggest one thing. Make a list of companies that produce competitive goods and services with your company. Ask yourself how you compare on these three factors—price, quality, and service. Taken as individual considerations they mean very little; however, when all three are present they represent an unbeatable combination.

THE ENGINEER VERSUS THE CHALLENGE

The industrial engineer is often in a position that is unique among the various engineering disciplines. He is often the one who coordinates the efforts of the other branches of Engineering to produce a product or service. He is also often the one who is in a position to coordinate the efforts of the production force with the goals of the business and strategies of upper management. It is this unique position that allows him to focus tremendous resources on the overall goal of productivity improvement. It is because of this and other unique skills that industrial engineers can be, and should be, one of the most effective productivity drivers of the business machine.

One key question to be asked at this pont is how are industrial engineers utilized in your company? Are they assigned to work in a narrow spectrum of the business? If this is the case, they are probably being underutilized. Management in most companies has developed a respect for the problem-solving attributes that most of these engineers utilize in their daily assignments.

Before discussing ways to increase productivity through industrial engineering, let us briefly review the relationship between productivity and inflation. If the cost of any of the inputs that go into a product or service is increased, the increase must be recovered by a corresponding increase in the price to maintain the economic health of the business. If the increase in price yields no increase in the intrinsic value of the product or service, inflation occurs. This price increase may, in turn, be waterfalled into other businesses and the inflationary cycle continued. If, on the other hand, the requirement for the resource can be reduced to an extent corresponding to the increased cost, there would be no need for a price increase. As an example, if wages increase by 10 percent but through improved work methods that product or service requires 10 percent less labor, there is no need for a price increase. Of course, the inflation problem is even worse, if, while the costs of our resources are increasing, the input of the resources is also increasing.

MEASUREMENT OF PRODUCTIVITY

What is this productivity factor that would seem to solve so many of our problems? Very simply stated, it is the ratio of our inputs to our outputs, the way we consume resources as related to what we produce.

Measuring productivity has never been and is not now an easy task. It has been done by using output per employee hour. Although not a very complete or accurate measure, this method is still used in many places, including government data. The assumption that all inputs are a function of labor becomes even less valid as industry becomes more technology intensive. A more progressive but exceedingly difficult measure would be to weigh the five factors of technology, human skill, capital, energy, and expense in proportion to their contribution to the output.

The leading contributor to increased productivity for the past few years has been technological progress. There is a direct relationship between technological progress and productivity increases. A further direct relationship may be seen between the amount of money allocated to research and development and technological progress. Although productivity increases may tend to stem primarily from the quality of technological improvements, an increased quantity of research and development does produce some increase in the quantity of innovations.

Let us review briefly a new technology being implemented within the Bell System—the transmission of information by a new development known as the Lightwave Transmission System. This system serves as the replacement for conventional copper or aluminum cable.

It carries telephone calls and other information in the form of pulsed bursts of laser-generated light over glass fiber cables that link telephone switching offices. Lightguide as a transmission medium offers the advantages of increased bandwidths, low loss, and small size.

From the viewpoint of productivity:

- The fiber itself is immune to electrical interference. This results in fewer resources required to shield the conductors and also a lower error rate in information transmitted.
- The Lightwave cable is smaller, lighter, and easier to install than wire cable. This is especially important in big cities where underground ducts are required, and those existing are already crowded.
- Each pair of hair-thin glass fibers will soon be capable of carrying up to 4000 times the information that can be sent over a pair of copper wires.
- The need for signal strength regeneration is reduced from every mile to every five miles.

In summary, Lightwave technology provides the ability for high increases in the information-carrying capacity over small, light cable. While it is difficult to equate the total productivity gains to the Bell System, in simple terms, we will soon be carrying over a ½-inch, 144-fiber cable what just a few years ago would have required 274,000 pairs of copper wires.[1]

RESEARCH AND DEVELOPMENT—A MUST

Recently though, on a national basis, the funds allocated to research and development have not been in proportion to the Gross National Product. The primary reason for this decrease has been due to fewer funds spent by the government for research and development. This fact, coupled with the many requirements placed upon industry to meet strict environmental and safety regulations, has directed much of the research and development money to areas other than those aimed at productivity improvements. An example of this specific problem is the auto industry. Much of their research and development money has been spent on ways to reduce pollution and to improve the safety of the cars. This research and development money was, therefore, not available to develop items such as new metalworking equipment to replace that which is existing. Over half of the U.S. metalworking equipment is over 20 years old. A comparison of the steel industries in Japan and the United States gives a good indication, for example, of what the lack of capital investment can do. In one Japanese plant, the steel-producing equipment was declared obsolete and replaced after 19 years. In this country, many plants are still using equipment that is 40 years old.

University research and development, a prime source of many innovations, has also been adversely affected by the lack of capital. In summary, it takes capital to turn technology into production facilities.

THE ENERGY CHALLENGE

In recent years energy has become an increasingly important part of the productivity equation. What was once plentiful and cheap, and virtually ignored in production planning, has become a primary consideration.

Comprehensive management-supported energy conservation is the key to an outstanding energy management program. At Western Electric the energy conservation program is a vital part of company policy. The program focuses on management/employee awareness programs, small technology innovations, as well as long-term, large capital investment projects. The program covers everything from large solar system installations to thermostat reductions. Energy usage is important to the company's conservation effort. About 40 percent of the energy is used to heat, cool, and light buildings. This percentage is relatively fixed at present. However, new buildings will impact heavily on this rate. The remaining 60 percent of the energy mix is used in manufacturing operations. Western Electric's energy conservation effort began in 1973 with the establishment of a corporate-level energy department and supported by personnel at each major manufacturing or service facility. This extensive energy management program has saved more than $177 million in fuel bills since 1973.

Western Electric's program consists of two phases: (1) administrative achievements and (2) technical achievements. The administrative phase is ongoing and includes educational, motivational, and energy awareness programs to keep energy conservation in the forefront of day-to-day activities.

The technical achievement phase of the program consists of four categories: building improvements, process improvements, new construction, and solar applications. All these factors affecting productivity are interrelated and must be considered as an entire system. In the same manner, productivity is not only a business or national problem but one of worldwide significance and one in which everyone must be concerned.

IMPROVING PRODUCTIVITY

Working to improve productivity must be done on a unified, coordinated, systematic, ongoing basis. A recent survey of corporate executives asking why productivity improvement programs fail brought the response from 70 percent that the main reason for failure was a piecemeal, or uncoordinated, approach. Goals for productivity improvement must be set and a specific plan for achieving these goals developed.

The process involved, be it manufacturing, distribution, or service, must be evaluated to define opportunities. The problem should then be analyzed to develop alternate solutions. As a final step, it should be improved by selecting and implementing the one offering the greatest return. Therefore, to achieve the same output with less input, or to use the same input for greater output, we must evaluate, analyze, and improve. The same old textbook approach we have been reading, studying, and talking about for years. Sounds simple, doesn't it? In actual practice it is a day-by-day challenge. However, the challenge when met head on will produce positive results. These steps were utilized for three related projects involving the distribution and warehouse operations at Western Electric's three major service facilities on the West Coast. We realized that we were running out of space. The provision of additional space would require the expenditure of extensive capital with no increase in the intrinsic value of the service—a classic case of a productivity drop. The situation was evaluated, and the major opportunities defined dealt with outmoded facilities and material allocation procedures. An analysis of the process was made, and specific areas in the distribution system were singled out for improvement. Various alternatives were then reviewed and the projects selected that would improve the distribution system while yielding the highest rate of return. Return on investment will continue to be a major consideration.

The projects, as they finally evolved, utilized a computerized material allocation system to lay out the storage facilities and assign material locations. With the addition of modern material-handling equipment, under the guidance of automated process controls, these projects resulted in an annual cost saving of over $2.5 million. These savings have been documented in our cost reduction programs.

This was achieved by working in a totally integrated mode utilizing a systems approach and bringing all involved organizations and people into the planning process. The operating people were able to contribute many excellent ideas pertaining to the detailed procedures of the new environment. In addition, their involvement in the project from the beginning made them much more receptive to the changes that were occurring in the work procedures.

This process of analyzing the operation to determine where the maximum opportunities lie was utilized very successfully in our warehouse projects. It was readily determined that about 20 percent of our items accounted for 80 percent of our volume. By concentrating our efforts on the items making up this 20 percent, we were able to achieve the greatest return on our investment for our efforts to improve the productivity of the warehouse and distribution operation.

The relocation of the pallet racks, a vital part of the rearrangement, demonstrates that the opportunity for productivity improvement exists in all areas. In the past, to relocate pallet racks the material had to be removed, stored temporarily, the racks disassembled, relocated, and reassembled, and the material returned to its designated position. A recent engineering development has alleviated the need for this very labor-intensive procedure. In the new method, pallets situated under the racks at floor level are moved. Air pallet jacks are connected to the racks. The racks, which are then floated on a thin cushion of air, can be readily moved by one person. After the racks are moved to their new location, the air pallet jacks are removed and the material replaced underneath. This procedure permits four times the number of racks to be relocated during a shift and at a considerable cost saving.[2]

In conclusion, without productivity increases we will continue to face higher inflation, increased unemployment, and, as a consequence, a lower standard of living. On the other hand, improved productivity is one means to a stronger economy with a decrease in inflation. It is important that we remain aware of the direct relationship between increased productivity and an improved economy for, as

Peter Drucker, noted business author and professor, has stated, "Productivity is the first test of management competence."

AN EXCELLENT EXAMPLE

The acid test for any cost reduction program is evaluation of the long-term price, quality, and service trends of a product or service. Many companies are engaged in the pursuit of cost reduction. The fruits of any successful program should be evident as you use the product or service. Most of us are prone to keep certain types of records—perhaps as a reference point at some future date. Some favorite items for me have been sticker prices from new car purchases. I started saving these items from a 1964 Pontiac station wagon, whose $3450 total price is now a fond memory. My next reference point, a 1970 Chevrolet station wagon with a sticker price of $4950, is very appealing when viewed in today's market. Things had certainly changed when the 1981 Chevrolet station wagons were introduced. Prices in the range of $12,900 were not uncommon. Needless to say, it was in this period that the term "sticker shock" was born.

Enough of the bad news. It would be a positive step to find something that could be used as a standard reference of days gone by that is still a good deal today. Such a product does exist and we all use it daily. The product is the telephone and the wide range of related services that we all have grown very dependent upon. What leads me to think this is such a good bargain in today's environment? My decision is based on three factors: price, quality, and service.

Refer to Figure 16-1. Compare the 1940 price data with the 1981 data. In today's environment what else can be cited as a comparable example? As a single statement for effective cost reduction this example speaks for itself.

Up to this point we have looked at various aspects of cost reduction effort. As we have discussed, our total interest has been directed at using a positive cost reduction program that can be translated into increased productivity. In order to make the conversion from the cost reduction mode to productivity improvement, we must be careful in dealing with various inputs. The following model example by B.W. Taylor and K.R. Davis provides an interesting insight that may fit your particular operating environment.

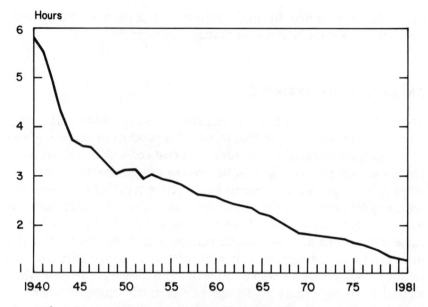

*One month's individual service with 100 local calls

Figure 16-1.   Cost of residential telephone service: hours of work required to pay for residential telephone service from 1940 to present. (Residential telephone service is one month's individual service with 100 local calls.) (*Source:* Western Electric Co.)

Use this model to measure and monitor changes in your firm's total productivity. All important input and output factors, including potentially troublesome raw materials and investor contribution inputs, are explained. Example demonstrates model's application.

Bernard W. Taylor, III, Virginia Polytechnic Institute and
State University, Blacksburg, Virginia
K. Roscoe Davis, University of Georgia, Athens, Georgia

CORPORATE PRODUCTIVITY—GETTING IT ALL TOGETHER*

Many corporate managers have expressed a desire for more comprehensive measures of companywide productivity. They recognize that such measures are useful in projecting costs and input requirements. However, much of the literature related to productivity has failed to expose managers and industrial engineers to the detailed analysis of "total factor" productivity measurement for a large manufacturing firm.

---

*Reprinted with permission from *Productivity: A Series from INDUSTRIAL ENGINEERING*. Copyright © American Institute of Industrial Engineers, Inc., 25 Technology Park/Atlanta, Norcross, Georgia 30092.

The Total Factor Model

Total factor productivity measures have been defined by several sources. (References 1, 2, 3, 4). These total models attempt to reflect all relevant output and input factors related to the manufacturing process, as opposed to more traditional partial measures such as labor productivity (5, 6). The total factor model used in this case analysis is formulated as:

$$TFP = \frac{(S + C + MP) - E}{(W + B) + [(K_w + K_f) \cdot F_b \cdot d_f]} \qquad (1)$$

Where,

S	=	sales		
C	=	inventory change	Total	
MP	=	manufacturing plant	Output	
E	=	exclusions		
W	=	wages and salaries	Labor	
B	=	benefits	Input	
K_w	=	working capital		
K_f	=	fixed capital	Adjusted	
F_b	=	investor contribution adjustment	Investor Input	
d_f	=	price deflator factor		

This model differs with other total factor measures in several ways, the most prominent being the exclusion of raw materials as an input. Many firms consider raw material purchases as the fruits of someone else's labor and, as such an obfuscation of one's own productivity effort. For this reason, raw materials have been excluded as an input in the model. However, some firms which have large material inputs do not feel this exclusion is justified. (To accommodate this view, we demonstrate use of this model both with and without material inputs.)

How It's Used

The example described is a major industrial firm. The period of analysis includes the years 1967 through 1973. The base year for the study is 1967. The initial adjusted productivity index for 1967 is thus 100. During the period of analysis the firm experienced a general trend of overall growth, with the exception of a temporary decline in 1969. To demonstrate the structure of the model, the computation of each component of output and input is analyzed.

Table I.   Net Adjusted Sales (millions of dollars).

	1967	1968	1969	1970	1971	1972	1973
Sales ($) for year	91.11	92.64	60.12	78.55	77.24	84.10	92.56
Price deflator[a]	100.0	86.4	78.8	70.3	62.7	57.3	51.4
Adjusted 1967 sales	91.11	107.22	76.30	111.74	123.19	146.77	180.08

[a]The price deflator experienced a rather rapid decline due to competitive market pressure.

Value Added Output.   It is that output which the firm contributes to the market value of its products. Total output consists of sales, plus an adjustment for inventory changes. Purchased materials and services and rentals represent output from other firms and thus are subtracted from total output.

Sales.   The starting point for computing total output was the firm's net sales billed for the years 1967 to 1973. Sales figures were deflated to 1967 dollars via an internal company price index. Table I is a summary of the sales figures.

Inventory Change.   When computing total output, the output value should reflect total production efforts of the firm. Therefore, changes in inventory must be included in the total output figure. If inventories have increased over a year's time, an addition to output would be warranted. If inventories have decreased, a deduction in output is necessary.

Inventories are of three types: raw material inventory, work-in-process inventory, and finished goods inventory. For the case example, the inventory information is summarized for the years 1967 to 1973 in Table II. When the ending inventory is subtracted from beginning inventory, the residue becomes the amount consumed or supplied annually for each type of inventory. Work-in-process inventories are divided into two parts, one-half raw materials and one-half finished goods (depending on their degree of completion). All values were deflated by the appropriate price index.

Manufactured Plant.   Manufactured plant includes items which could be purchased from outside sources but are produced internally. These include

Table II.   Change in Inventory (millions of 1967 dollars).

	1967	1968	1969	1970	1971	1972	1973
Raw materials	3.41	(0.42)	(0.36)	1.73	1.22	2.40	4.78
Finished goods	2.01	(0.05)	0.40	0.96	0.33	0.71	1.02
½ Work-in-process (raw material)	0.96	0.51	0.21	1.01	2.76	3.11	3.53
½ Work-in-process (finished goods)	0.58	0.30	0.32	0.69	(0.53)	(0.01)	0.76
Total inventory change	6.96	0.34	0.57	4.39	3.78	6.21	9.33

such items as internal maintenance and repairs, internally-produced machinery and equipment, and research and development. Since labor (in terms of salaries and wages) is included as a component of total input, the total results of that input must be recognized in terms of its contribution to output; thus, internal maintenance and repairs and internally-produced machinery and equipment, while not contributing directly to product output, must be included as an output.

The inclusion of manufactured plant is often neglected in measuring productivity. This usually occurs when a firm is small and provides a minimum of its own plant and maintenance. The firm in this example was quite large and produced approximately $5 million yearly in manufactured plant. Depending on the size of the firm and the extent of its operations, this output factor may be deleted.

Research and development (R&D) is often excluded as an output factor, primarily because of the difficulty in estimating its value. However, in some industries research and development is of such magnitude that its exclusion could result in a gross error in the overall productivity index. R&D, as an output factor, is defined as total expense for technical effects in long-range research and development efforts.

For the example case, R&D, as well as other manufacturing plant factors, was adjusted via appropriate price and cost indices to reflect 1967 dollars.

Exclusions. Factors which do not represent results of production efforts must be subtracted from total gross output. These factors include externally-purchased materials and supplies (M&S), depreciation on buildings, machinery and equipment, and rentals. For the example case, each of these yearly values were computed and then deflated to reflect 1967 base prices. Direct materials were deflated by the wholesale price index.

The exclusion of raw materials from output and input is the major difference of a value-added approach compared to other total productivity models. The argument generally made is that the quality of raw materials can influence the amount of output from a firm. In addition, however, the argument can easily be made that raw materials do not realistically reflect a firm's technological progress; raw materials reflect the progress and efficiency of an external operation. Raw materials, thus, are excluded.

Depreciation was treated as an exclusion since it is similar to a time payment for machinery or plant purchased in external segments of the economy. Depreciation charges do not reflect actual output from capital and labor inputs.

All exclusions are summarized in Table III along with total output. The results of Table III yield total "value-added" output, the numerator of equation (1).

Table III.   Total "Value Added" Output (millions of 1967 dollars).

	1967	1968	1969	1970	1971	1972	1973
Gross output	103.46	112.34	81.40	121.89	133.68	157.36	194.72
Sales	91.11	107.22	76.30	111.74	123.19	146.77	180.08
Manufacturing plant	5.39	4.78	4.53	5.76	6.71	4.38	5.31
Inventory change	6.96	0.34	0.57	4.39	3.78	6.21	9.33
Total exclusions	63.33	65.16	47.26	72.31	80.87	97.12	123.39
M & S	59.65	61.05	43.50	66.30	72.22	85.74	107.21
Depreciation	3.12	3.53	3.59	4.78	6.21	7.83	10.17
Rentals	0.56	0.48	0.17	1.23	2.44	3.55	6.01
Total "value added" output	40.13	47.18	34.24	49.58	52.81	60.24	71.33

Total Inputs.   Inputs of the productivity model include the two key factors of production, capital and labor. Most productivity models include only labor inputs because of difficulties associated with computing capital input measurements. However, if one is to have a true total factor productivity model, a means must be provided for including the yearly contribution of fixed and working capital to the production of goods.

Labor Input.   Labor input includes all monetary compensation paid to hourly and salaried employees, including all benefits. Total annual compensation includes such items as overtime, vacation, sickness, insurance, profit sharing, social security tax, retirement, bonuses, etc. For the example case, the total annual values were determined and then adjusted for 1967 dollars. Total labor input is summarized in Table IV.

Capital Input.   The capital of a company is composed of tangible assets (buildings, machinery and equipment, and inventories) and intangible assets (cash, notes, accounts receivable, portfolio investments). Except for portfolio investments, both tangible and intangible assets are used either directly or indirectly in the production of output.

Table IV.   Total Labor Input (millions of dollars).

	1967	1968	1969	1970	1971	1972	1973
Total labor Compensation	34.17	35.63	31.49	36.85	38.94	41.14	46.78
Wages & salaries	25.63	24.94	24.25	25.43	28.04	28.39	33.21
Benefits	8.54	10.69	7.24	11.42	10.90	17.75	13.57
Consumer index	100.00	101.80	103.50	105.20	107.00	108.90	110.90
Total labor (1967 dollars)	34.17	35.00	30.42	35.03	36.39	37.78	42.18

Table V.   Total Capital Assets (millions of 1967 dollars).

	1967	1968	1969	1970	1971	1972	1973
Working capital	26.14	34.78	36.01	50.24	56.60	62.13	71.08
Fixed capital	41.13	44.38	45.61	48.22	48.22	57.91	63.07
Total capital assets	67.27	79.16	81.62	99.06	109.30	120.04	134.15

Capital inputs can generally be obtained from the company balance sheet. Capital is divided into two parts: fixed capital and working capital. Working capital includes cash, notes, accounts receivable and inventories, and prepaid expenses. Fixed capital includes land, building, machinery and equipment, and deferred charges.

Annual working and fixed capital for the example is summarized in Table V. All values were adjusted to 1967 dollars.

Investor Contribution.   One of the most difficult aspects of developing a total factor productivity measure is annualizing total capital. We use a rather simplified approach that we label the "investor contribution" approach. Investor contribution is defined as real net capital (i.e., capital after depreciation) for each year weighted by the rate of return in the base year of 1967.

To determine the annual value for the example case, total capital assets were computed for every year (as noted in Table V). Profit before taxes was then computed as a percent of total capital assets for the 1967 base year. This figure was then used as the rate of return on capital for the base year. Using this percentage as a constant, applied against all succeeding years, the investor contribution was computed for each year.

In 1967, the investor contribution was roughly 17 percent of total assets. This percentage was used to compute investor contribution for the remaining years through 1973. Investor contribution was adjusted for 1967 prices via a GNP price deflator. Computations are summarized in Table VI.

Table VI.   Investor Contribution (millions of dollars).

	1967	1968	1969	1970	1971	1972	1973
Total capital assets	67.27	79.16	81.62	99.06	109.30	120.04	134.15
Profit before tax[a]	11.86						
Profit as % of assets	17.60	17.60	17.60	17.60	17.60	17.60	17.60
Investor contribution	11.86	13.93	14.37	17.43	19.24	21.13	23.61
Price adjuster	100.00	101.60	103.50	106.10	108.00	109.80	112.80
Investor contribution (1967 dollars)	11.86	13.71	13.89	16.43	17.81	19.24	20.93

[a] An adjustment for depreciation charges was made in this value.

Table VII.   Total Inputs (millions of dollars).

	1967	1968	1969	1970	1971	1972	1973
Total gross assets	67.27	79.16	81.62	99.06	109.30	120.04	134.15
Working capital	26.14	34.78	36.01	50.24	56.60	62.13	71.08
Fixed capital	41.13	44.38	45.61	48.22	52.70	57.91	63.07
Total capital input	11.86	13.71	13.89	16.43	17.81	19.24	20.93
Total labor input	34.17	35.00	30.42	35.03	36.39	37.78	42.18
Total input	46.03	48.71	44.31	51.46	54.20	57.02	63.11

The combined input for the total factor productivity model is the sum of labor and capital outlays adjusted to 1967 dollars. This total is shown in Table VII.

Results Without Raw Materials

This analysis demonstrates the methodology for computing the input and output factors necessary for the total factor model. All factors have been expressed in 1967 base period prices. In 1967, the productivity index is equal to 100.00; all other years are computed in relation to the base period. The results are summarized in Table VIII. This table includes partial indices for capital and labor as well as the total factor index. Notice that in 1969 productivity dropped severely due primarily to severe price pressure in the industry. However, productivity jumped substantially in 1970, and the firm experienced a steady growth thereafter.

The data shown in Table VIII, particularly the last line of the table, indicates that the firm experienced significant changes in productivity over the

Table VIII.   Productivity Indices—Total Factor Productivity Model (millions of dollars).

	1967	1968	1969	1970	1971	1972	1973
Output	40.13	47.18	34.24	49.58	52.81	60.24	71.33
Total output	46.03	48.71	44.31	51.46	54.20	57.02	63.11
Labor	34.17	35.00	30.42	35.03	36.39	37.78	42.18
Capital	11.86	13.71	13.89	16.43	17.81	19.24	20.93
Output labor	1.174	1.348	1.125	1.415	1.451	1.594	1.691
Labor index	100.00	114.82	95.83	120.53	123.59	135.78	144.03
Output capital	3.383	3.441	2.465	3.017	2.965	3.131	3.408
Capital index	100.00	101.71	72.86	89.18	87.64	92.55	100.73
Total index	0.871	0.969	0.772	0.963	.974	1.056	1.130
Adj. productivity index	100.00	111.25	88.63	110.56	111.82	121.23	129.73
Precent change from previous year	—	+11.25	−20.33	+24.74	+1.14	+8.42	+7.01

Table IX.   Productivity Indices—All Inclusive Model (millions of dollars).

	1967	1968	1969	1970	1971	1972	1973
Output	103.46	112.34	81.40	121.89	133.68	157.36	194.72
Total input	109.36	113.87	91.57	123.77	134.57	154.14	186.50
Labor	34.17	35.00	30.42	35.03	36.39	37.28	42.18
Capital	11.86	13.71	13.89	16.43	17.81	19.24	20.93
Materials & others	63.33	65.16	47.26	72.31	80.87	97.12	123.39
Output/labor	3.02	3.20	2.67	3.47	3.67	4.16	4.61
Labor index	100.00	105.96	88.41	114.90	121.52	137.75	152.65
Output/capital	8.72	8.19	5.86	7.42	7.51	8.18	9.30
Capital index	100.00	93.92	67.20	85.09	86.12	93.81	106.65
Output/materials & others	1.63	1.72	1.72	1.69	1.65	1.62	1.58
Material index	100.00	105.52	105.52	103.68	101.23	99.38	96.93
Output/ Total input	0.946	0.986	0.889	0.985	0.993	1.021	1.044
Total index	100.00	104.23	93.97	104.12	104.97	107.93	110.36
Percent change from previous year	—	+ 4.23	−9.84	+ 10.80	+ 0.82	+ 2.82	+ 2.25

seven-year period. An examination of the labor index and capital index indicates that the productivity resulted somewhat from increased labor performance. But the indices could be interpreted thus: The industry became more capital based, therefore total productivity increased because of more automated systems—the capital index declined because of large capital input—that resulted in increased labor performance.

Results with Raw Materials

The total factor model developed here excluded raw materials as a productivity input. Some would argue that this is a significant and undesirable deletion. Many researchers have indicated that materials and their handling and usage must be considered as an important component of any productivity measure (7, 8).

If one were to modify the total factor productivity model by adding previously excluded items (materials and supplies, depreciation and rentals) to both output and input, a modified productivity performance would result. Table IX is a summary of the computations for an "all-inclusive" model. The key difference between Tables VIII and IX is the inclusion of "Materials and Others" under Total Input.

An examination of the last line in Table IX shows that the firm still experienced some significant changes in productivity over the seven-year

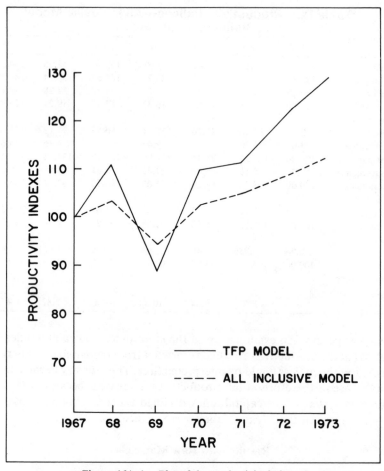

Figure 16A-1. Plot of the productivity index.

period, but the magnitude of the change from year to year was less than that in Table VIII. The impact of including "materials and others" in the model thus dampened the productivity change (both positive and negative) of the firm. This resulted, primarily, because of the size of the "materials and others" input compared with labor and capital inputs, Figure 1.

Postscript

The best productivity model, of course, has yet to be written. Our model may not meet the needs of a particular firm; however, because of its general nature, it can be easily modified to reflect the needs of a specific company.

References

1. Craig, Charles E. and Harris, Clark R., "Total Productivity Measurement at the Firm Level," *Sloan Management Review,* Spring, 1973, 13–29.
2. Hines, W.W., "Guidelines for Implementing Productivity Measurement," *Industrial Engineering,* June 1976, 40–43.
3. Kendrick, J.W. and Creamer, D., *Measuring Company Productivity,* Studies in Business Economics, no. 89, National Industries Conference Board, New York, 1965.
4. Mundel, M.E., "Measures of Productivity," *Industrial Engineering,* May 1976, 24–26.
5. Greenberg, Leon, *A Practical Guide to Productivity Measurement,* The Bureau of National Affairs, Inc., Washington, D.C., 1973.
6. Law, Donald E., "Measuring Productivity," *Financial Executive,* October 1972, 24–27.
7. Reed, R. "Material Handling Cost Factors," *Industrial Engineering,* March 1976, 30–36.
8. Smith, J.F., "Productivity Through Material Handling," *Industrial Engineering,* March 1976, 20–21.

REFERENCES

1. Criner, E.A., Frost, S.N., and Mahler, E., AIIE Region X, Annual Conference, "Productivity Drivers," October 1981.
2. Ibid.

Index

GLASSBORO STATE COLLEGE